SCIENTIFIC ANALYSIS ON THE POCKET CALCULATOR

SCIENTIFIC ANALYSIS
ON THE
POCKET CALCULATOR

JON M. SMITH
SOFTWARE RESEARCH CORPORATION

A WILEY-INTERSCIENCE PUBLICATION

JOHN WILEY & SONS
New York . London . Sydney . Toronto

Library of Congress Cataloging in Publication Data:

Smith, Jon M   1938–
    Scientific analysis on the pocket calculator.

    "A Wiley-Interscience publication."
    Includes index.
    1. Calculating-machines.   2.   Numerical analysis.
I.   Title.

QA75.S555   510'.28   74-20713
ISBN 0-471-79997-1

Printed in the United States of America

10 9 8 7 6 5 4

To Laurie,
Mike, and Chris

PREFACE

This book is written for all those who own or operate a modern electronic pocket or desk calculator, and especially engineers, scientists, science students, mathematicians, statisticians, physicists, chemists, computer analysts, and science educators.

When the right numerical methods are used, the electronic pocket calculator becomes a very powerful computing instrument. "Micronumerical methods" that will help the reader to derive the most computing capability for every dollar he has spent on his pocket calculator are discussed here.

Most of the methods work on *any* pocket calculator. Special methods for certain types of machines are clearly indicated where necessary. Key stroke sequences for both algebraic and reverse-polish calculators are shown. Virtually all pocket calculator keyboards and capabilities were considered in preparing this book, to ensure that the numerical methods presented are the most universally applicable for general pocket calculator analysis.

Each part of this book provides a consistent and careful treatment of the methods and tabulated formulas that can be used with a pocket calculator. The aim is to supply the reader with a large number of numerical techniques, numerical approximations, tables, useful graphs, and flow charts for performing quick and accurate calculations with pocket calculators.

The numerical methods are presented from four viewpoints:

1. The numerical evaluation aspect of each numerical method.
2. The manner in which each method is used.
3. The limitations and advantages of each method.
4. The tabulation of useful formulas in forms that are most convenient for pocket calculator analysis.

Emphasis is also given to numerical methods used in certain types of data processing, such as harmonic and statistical analysis. And they are presented in forms that are directly useful to engineers, scientists, and programmers.

The premise of this book is that the pocket calculator provides the scientific analyst with an important new dimension in analysis. Obviously the pocket calculator is useful both for numerically evaluating functions and for processing data. *In addition,* it enables the analyst to quickly gain detailed and quantified knowledge about any technical discipline (his own or another's) by *learning* its mathematical models and tools through use and experimentation on the pocket calculator. In short, the pocket calculator becomes a *learning machine* for the scientific analyst. A scientific analyst no longer need first develop a mathematical model for a complex process or system being studied and then turn it over to a computer programmer for its numerical evaluation. Instead, he can numerically evaluate complex functions (and thus analyze complex problems) on even the simplest four-function calculators in his home or office.

Finally, throughout the world scientific analysts working on pocket calculators are inventing their own numerical methods for evaluating problems in their specific disciplines. In this sense, the pocket calculator is a *research tool* which the analyst can use to develop his own numerical methods for his own purposes.

Throughout the book I give more attention to subjects of interest to the practitioner than to those of interest to the theorist. Though the treatment of this material is mathematical, I have not strived for conciseness or rigor beyond that required for pocket calculator analysis. Numerous examples of each technique and method are given, and their implementation is discussed in detail.

This book consists of four parts that are subdivided into 12 chapters, each dealing with topics in numerical analysis that are useful to the practical analyst. I have tried to avoid overgeneralization in the treatment of these topics, since numerical analysis is an art as much as it is a science. *Part I* of the book introduces the spectrum of pocket calculators (including their capabilities and their limitations) available to engineers and scientists. Particular attention is given to the unique computing features of interest to the scientific analyst. Part I also presents mathematical preliminaries and mathematical refresher material and develops certain elementary numerical methods particularly suited to analysis on the pocket calculator. Topics from arithmetic to algebra and analysis with complex variables are covered.

Part II presents numerical methods and formulas for numerically evaluating advanced mathematical functions. It also deals with the nested parenthetical form of the most frequently used functions in advanced engineering mathematics. It is the nesting of a sequence of arithmetic operations in parenthetical form that is the basis for performing advanced analysis on the pocket calculator. For example, 14 multiplies, 2 divides, 2 sums, and 108 data entries, totaling 126 key strokes and 5 data storage records, are needed for a three-digit floating-point evaluation of sin (x) $\approx x - x^3/3! + x^5/5!$. But only 54 key strokes and *no data storage records* (on a scratch pad) are needed to evaluate sin $(x) \approx x(1 - (x^2/6)(1 - x^2/20))$ to the same accuracy. Though we would evaluate sin (x) in this manner only on a four-function calculator, this example does illustrate the point that many complex formulas usually requiring calculator memory to be numerically evaluated can be written in a "nested" form not requiring calculator memory and thus can be evaluated conveniently on even the simplest four-function pocket calculator.

The nested parenthetical form is considered to be a "fast" form for numerical evaluation. That is, functions written in nested parenthetical forms require fewer operations to numerically evaluate than do the same functions in their "simplest algebraic form." The nested forms are therefore evaluated more rapidly than their unnested counterparts.

In Part II the topics of recurrence formulas for evaluating advanced functions and performing analysis are also covered. Recurrence formulas are unique in that they are *infinite memory forms* of otherwise *finite memory form* calculations. The formulas give the pocket calculator "virtually" an infinite memory for storing data, which creates many numerical methods for data processing (as in statistics) to be rewritten in recursive form for pocket calculator analysis. Here, again, the emphasis is on making even the simplest four-function calculator capable of doing sophisticated analysis without memory. Such concepts as nested parenthetical forms and recursion formulas, when combined with those of Chebyshev economization and rational polynomial approximation, provide tremendous flexibility and accuracy in the numerical evaluation of even the most complex functions on the simplest four-function pocket calculator. In fact, the serious analyst can perform precision calculations unheard of until a few years ago—in the comfort and convenience of his home or while traveling on the job.

Part III examines the methods and formulas for performing advanced "types" of analysis on the pocket calculator. Included are such topics as numerical evaluation of definite integrals and methods for numerical differentiation of data sets, solving differential equations, simulating linear processes, performing statistical analysis, and performing harmonic analysis.

Part IV deals exclusively with analysis on the advanced programmable pocket calculator. The chapters illustrate conclusively the leap in computing capability produced by the pocket calculator. They are based on personal experience in solving a very large number of problems on the programmable pocket calculator developed by the Hewlett-Packard Corporation. The discussion is general, however, recognizing that more programmable pocket-style calculators are being developed.

This book grew out of eight years of study on numerical methods for analysis on the digital computer. These methods were revised over a period of three years to make them applicable to desk calculator analysis and eventually to pocket calculator analysis. A number of the methods have been available to the analyst in scattered literature, such as user's guides and manuals for desk-top and pocket calculators, journal articles, and some textbooks. A large part of the material was developed by the author or was provided by my associates in industry. I am particularly indebted to my associates, at the Software Research Corporation and the McDonnell-Douglas Corporation. They generously shared with me many of their "tricks of the trade" and suggested interesting problems for this book. I express my sincere appreciation to one of the great numerical analysts of our time, Dr. Richard Hamming, of Bell Laboratories, for his review and improvements to the manuscript.

My thanks to the people at Hewlett-Packard who reviewed and critiqued the manuscript, and in particular to the HP-65 chief engineer, Mr. Chung Tung.

I want to thank Joseph and Sara Goldstein, who taught me the Goldstein algorithm—"one at a time."

To my wife, Laurie, my special appreciation for putting up with the 4 a.m. Writing schedule.

Finally, my thanks go to Mrs. Florence Piaget who typed the manuscript and helped me to prepare it for publication.

JON M. SMITH

St. Louis, Missouri
August 1974

CONTENTS

PART IV   THE PROGRAMMABLE POCKET CALCULATOR

PART ONE

INTRODUCTION TO POCKET CALCULATOR ANALYSIS

CHAPTER 1

THE POCKET CALCULATOR

1.1 INTRODUCTION

This chapter discusses the mathematical differences among the various pocket calculators and certain mathematical concepts, useful for analysis on the pocket calculator, that appear throughout this book.

We are not so much concerned with the hardware implementation of mathematical operands and operations as with the different ways in which they can be assembled in a computing machine—the hardware architecture. Only the most obvious mathematical aspects of calculator design are examined, such as the language used, the size and type of memory, the instruction set, type of input/output, and whether the calculator is programmable. There are some 432 types of calculators that could be hardware implemented. An entire book could be written on this subject alone. Here we limit our discussion to the more important mathematical differences that result from the various hardware implementations in order to:

1. Understand pocket calculators and the organization of mathematics within them.

2. Determine, in a cursory way, the combinations of hardware implementation that result in a significant jump in calculating capability.

The purpose is to narrow the types of calculator to be considered in this book to three.

Three hypothetical calculators that are typical of the available and anticipated pocket computing machines are discussed. Care has been taken throughout not to limit the methods of analysis to any particular hardware implementation. In fact, if there is bias throughout the writing it is in the direction of anticipated developments in the pocket calculator field, though its overall effect on the material is negligible.

3

The following mathematical aspects are covered in this chapter:

1. Arithmetic calculations.
2. Function evaluation with and without memory.
3. Computational accuracy.

The first is a thorough introduction to what appears to be a mundane subject (arithmetic on the pocket calculator). In fact, it is found to be quite the opposite because the different languages used by different calculators lead to significantly different capability for handling complex problems.

Particular attention is given to nested parenthetical forms of functions that permit function evaluation on memoryless and limited-memory calculators. Nested parenthetical forms are used as a means of providing implicit memory to the memoryless calculators. They are also "fast" in the sense that their evaluation involves fewer key strokes than the usual algebraic form.

No chapter on mathematical preliminaries in a book on numerical analysis would be complete without a discussion of computational accuracy. Here we examine:

1. The accuracy limitations of the typical pocket calculator.
2. Ways in which to accurately evaluate functions in general, and on the pocket calculator in particular.

1.2 MATHEMATICAL DIFFERENCES IN POCKET CALCULATORS

Today's pocket calculators differ mathematically in many ways. Only the six more commonly encountered mathematical distinctions are covered here. In a sense, these are the major distinctions because they are the fundamental issues addressed in the *conceptual design* of every pocket calculator. The important mathematical distinctions that are associated with the subtleties of *detailed design* are not discussed because the hardware implementations vary widely. Perhaps the best known difference is that between the use of fixed-point and floating-point numbers.

The *fixed-point numbers* are those whose decimal point is fixed by the electronic circuitry. The difficulty is that when multiplying two large numbers together, so that the most significant digit exceeds the size of the numeric display, the number is truncated not in the least significant digits but in the most significant digits. Most fixed-point arithmetic computers have a symbol that is illuminated to indicate the *overflow* condition.

Floating-point numbers have a decimal point that moves so as to retain the most significant digits in any calculation. When a number is computed that is larger than the calculator's field of numbers and the decimal point location is unknown, most calculators display the most significant digits and illuminate a symbol indicating that the decimal point location is unknown.

In these two number systems, it should be noted, the number fields are dramatically different. In the floating-point number system the numbers are "bunched" around zero. In the fixed-point number system the numbers are uniformly distributed over the range of the number field. To see this, consider the process of incrementing each of these types of numbers on a pocket calculator.

The smallest possible increment between any two numbers is the least significant digit in the numeric display. For an eight-digit display with a decimal point fixed in the third place, the smallest increment that can be added to *any digit* is 0.001. Now consider the addition of an increment to a floating-point number. Since the decimal point "floats" in the floating-point number system, the decimal point precedes the far-left digit. For an eight-digit display, the smallest number that can be added to *zero* in a floating-point number system is 0.00000001. Now consider incrementing a floating-point number when the decimal point is after the far-right digit. In this case, the smallest number that can be added to 99999998 is 1. The difference in the size of the "smallest number" when incrementing a full and empty register in floating-point numbers is a factor of 10^8.

Now consider the full range of the positive numbers in both number systems. The fixed-point numbers range from 0.001 to 99999.999. The difference between numbers, no matter where a number is over the range of the calculator, is 0.001. Thus the numbers over the range of fixed-point numbers are uniformly distributed.

Again consider the range of the positive floating-point numbers, from 99999999 to 0.00000001. Clearly, the range is greater in the floating-point number system than in the fixed-point number system, but note also that when the numbers are very small the distance between them is 0.00000001. When the register is full, the difference between the numbers is 1. Obviously, over the range of floating-point numbers, the distribution is not uniform. In fact, there are as many numbers grouped between 0 and 1 as there are between 1 and the full register size 99999999.

It follows, then, that in fixed-point arithmetic the absolute difference remains fixed over the entire range of the number system, while in floating-point numbers the absolute difference varies significantly. It is worth emphasizing that in floating-point arithmetic the percentage difference remains fixed, while in the fixed-point arithmetic system the constant difference remains fixed over the range of numbers. As used here, *percentage difference* is the ratio of the difference between two consecutive numbers divided by the larger of the two. For most engineering analysis, percentage difference and percentage error are usually the measure of accuracy of most interest.

The floating-point number system is usually extended by powers of 10, permitting the positive floating-point numbers to range from 10^{-99} to 99999999×10^{99}. In fact, calculators are usually configured to display this extended number field in scientific notation. Interestingly, this even further bunches the floating-point numbers in the neighborhood of zero. Because of this grouping property of the floating-point numbers, the absolute errors are smaller for calculations with numbers between 0 and 1 than for numbers betwen 1 and the full range of the calculator.

From a hardware architecture viewpoint, fixed-point numbers are usually displayed with greater accuracy than floating-point numbers; and floating-point numbers are usually displayed with a greater dynamic range than fixed-point numbers. This can be seen by considering a register with eight *display elements* where we configure both fixed-and floating-point numbers. In fixed-point arithmetic, eight mantissa digits can be displayed. If the decimal point is allowed to be set by the decimal point key $\boxed{\,\cdot\,}$, and a display element is used to show the decimal point, then only 7 digits remain to display the mantissa. If scientific notation is used to increase the dynamic range of the display, $m+1$ display elements are required to display m digits in the exponent. The extra display element is used to show the sign of the exponent.

Power of 10	Display	Required Display Elements
$10^{\pm x}$	$(\pm)(x)$	2
$10^{\pm xx}$	$(\pm)(x)(x)$	3
.		
.		
.		
$10^{\pm xxx}$	$(\pm)(x)(x)\cdots(x)$	$m+1$

$$\underbrace{}_{m \text{ digits}}$$

If 99 orders of magnitude are to be shown in the display register (the usual case with scientific pocket calculators) three display elements are required to display the exponent and its sign, leaving only five digits for displaying the mantissa. In this sense, then, the effect of increasing the display's dynamic range is to reduce the number of digits for displaying a mantissa, thus reducing the accuracy with which a number can be displayed.

1.3 INSTRUCTION AND DATA ENTRY METHODS

We discuss three types of data entry methods (languages) commonly used in pocket calculators: *polish, reverse-polish,* and *algebraic.* In polish notation, the *operator* precedes the *operand.* For example, to instruct the calculator to add the numbers A and B, in the polish entry method we would stroke the plus key, then enter the two numbers A and B. The logical operation in the machine would then display the result without the need for striking an additional key. In reverse-polish, the process is reversed; that is, the operands are introduced before the operator. In algebraic notation, the operator is sandwiched between the two operands. If we compute the sum of A and B in algebraic notation, we first input A, then stroke the summation key, follow that with an input of B, followed by stroking the equal key, whereupon C would be displayed in the register. It might seem that one entry method would result in many fewer key strokes than another entry method when numerically evaluating a function, but it turns out that the key strokes associated with instructing the calculator are fairly small compared with those associated with data entry. Far more important is the fact that certain entry methods, when combined with memory, result in the need for fewer data inputs or "scratch-pad" storage. The most common entry methods used in pocket calculators are the reverse-polish and algebraic methods, the former usually being used with machines that have a *memory stack* and the latter being attractive because of its "natural" algebraic treatment of numerically evaluating algebraic functions.

The natural way in which the algebraic method is used to numerically evaluate algebraic functions can be seen in the following example. Consider the relation

$$A \times B + C = Y$$

When evaluated on an algebraic notation pocket calculator (such as the Texas Instruments SR-10), the sequence of key strokes is*

$$\text{CL } A \times B + C = xxxxxxxx \; xx$$

The same function evaluated on a reverse-polish notation calculator (such

---

*Here the symbols CL, $\uparrow$, $\times$, $+$, and $=$ mean, respectively, "to clear the display register," "to store what is in the register in a temporary location," "to multiply," "to add," and "to present the answer—Algebraic language only."

as the Hewlett-Packard-35) would involve the sequence of key strokes

$$CL \ A \uparrow B \times C + xxxxxx \ xx$$

It is apparent that the former is more natural for simple functions than the latter. Reverse-polish notation, when used in conjunction with memory stacks, has the convenient property that it easily implements the numerical evaluation of functions with parenthetical expressions. This is not the case with algebraic notation. For example, the sum of products

$$(A \times B) + (C \times D)$$

must be rewritten in the form

$$\left(\frac{A \times B}{D} + C\right)D$$

to be evaluated using algebraic notation without using a scratch pad. The key strokes and operation to evaluate the sum of products *directly* is

$$CL \ A \times B = xxxx \ xx \ \text{STORE ON SCRATCH PAD} \urcorner$$

$$CL \ C \times D = yyyy \ yy + \text{INPUT} \ xxxx \ xx = zzzz \ zz$$

The key strokes to evaluate this sum of products in the rewritten form is

$$CL \ A \times B \div D + C \times D = zzzz \ zz$$

The reverse-polish with stacks evaluates the sum of products conveniently with key strokes

$$CL \ A \uparrow B \times C \uparrow D \times + zzzz \ zz$$

To avoid rewriting expressions in somewhat unfamiliar forms, the algebraic programming language can be designed to recognize a hierarchy among the operators, that is, when products are computed before sums or vice versa. The algebraic method with a "product-before-sum" hierarchy (such as the Texas Instruments SR-50) would evaluate the sum of products directly with the following key strokes:

$$CL \ A \times B + C \times D = zzzz \ zz$$

It has difficulty, however, with the expressions of the product of sums

$$(A + B) \times (C + D)$$

in that the hierarchy is set up to "multiply-before-add" rather than "add-before-multiply" which the product of sums requires. This problem is resolved with an additional storage location in which to store the intermediate sum. The key strokes are then,

$$\text{CL } A + B = \text{STO CL } C + D \times \text{RCL} = zzzz\ zz$$

In reverse-polish with stacks, the key strokes are

$$\text{CL } A\uparrow B + C\uparrow D + \times\ zzzz\ zz$$

Here STO means "store in memory" and RCL means "recall from memory."

1.4 MEMORIES

Pocket calculators are available with no memory, memory for a constant term, a memory stack of three to four registers, and in the more sophisticated machines addressable memory registers as well as the stack. The pocket calculator with a memory that simply retains a constant is characterized by the rather inflexible storage of a constant number that can be recalled or not recalled to the display register, as the operator desires. The stored constant can be used as a coefficient in multiple products or as a constant in multiple sums. The constant memory register does not automatically interact with the display register in most pocket calculators.

Pocket calculators with *memory stacks* generally involve three or four registers that can be manually *"pushed up"* and automatically *"pushed down"* for the purpose of retaining numbers developed in the display register. When used in conjunction with reverse-polish notation, they provide the first quantum level of computing capability above that present in the simple four-function memoryless pocket calculator. Data are usually entered into a stack with an entry operation. The three stack registers of a Reverse-Polish machine can be filled with three different numbers and then, as the operation on the number in the display register and the bottommost number in the stack is called for, the result is displayed in the display register and the stack automatically moves down, bringing the second number in the stack now to the first number, and the third number to the second. This process can be continued until the stack is empty. The algebraic machine with hierarchy uses the stacks somewhat differently. When a key stroke sequence is to be evaluated (upon key stroking the equal sign), the calculator first looks for products to evaluate and put into the memory stack and then executes sums (in a "multiply-before-add" hierarchy). The stack manipulations are automatic.

In the more sophisticated calculators with addressable registers, the process of storing data in a register is similar to that of storing data in the memory of a computer. The addressable stack does not interact unless programmed to do so. This memory resembles the register for storing a constant in the simpler calculators. Here, however, the ability to address the register is involved.

1.5 INSTRUCTION SET

The basic "four-function" calculator has keys for instructing the calculator to add, substract, multiply, and divide. What is amazing is that these small four-function machines, purchased at relatively low cost, can provide tremendous computing power. Examples of the use of the four-function pocket calculator for evaluating some of the most sophisticated engineering analysis will be seen later. Another arithmetic operation that can be performed with the four-function machine is computing powers of a given variable through repetitive multiply operations. While squaring a number involves only two multiplies, the number must be double entered. Thus, the simplest additional instruction that can be added to a pocket calculator that reduces the number of key strokes is the squaring operation or modifying the multiply instruction to square a number when only one data entry has been made.

Entirely new capabilities are added when the square root and reciprocal instructions are added to the calculator instruction set. There is no single stroke way on a four-function calculator to numerically invert a number without using a scratch pad and double data entry.* A similar situation holds for the square root. Thus we find the next most sophisticated pocket calculator to be a seven-function calculator, including square, square root, and reciprocal functions implementable with a single key stroke. Beyond this, additional instructions are added to aid in special-purpose computing in a variety of ways. The underlying theme throughout the addition of functions to a pocket calculator keyboard is to reduce the number of key strokes associated with data inputs.

Because we will be continually referring to instructions found on most scientific calculators, let us define the instruction sets that we use in this book:

*See Appendix A1-5.

Key Symbol	Key Name	Key Instruction
CL	Clear	Clears information in the calculator and display and sets the calculator at zero
0 1 ⋯ 9	Digit	Enter numbers 0 through 9 to a limit of an eight-digit mantissa and a two-digit exponent
·	Decimal point	Enters a decimal point
EE	Enter exponent	Instructs the calculator that the subsequent number is to be entered as an exponent of 10
CHS	Change sign	Instructs the calculator to change the sign of the mantissa or exponent appearing in the display
+	Add	Instructs the calculator to add
−	Subtract	Instructs the calculator to subtract
×	Multiply	Instructs the calculator to multiply
÷	Divide	Instructs the calculator to divide
x^2	Square	Instructs the calculator to find the square of the number displayed
$\sqrt{x}$	Square root	Instructs the calculator to find the square root of the number displayed

Key Symbol	Key Name	Key Instruction
$1/x$	Reciprocal	Instructs the calculator to find the reciprocal of the number displayed
sin	Sine	Instructs the calculator to determine the sine of the displayed angle
cos	Cosine	Instructs the calculator to determine the cosine of the displayed angle
tan	Tangent	Instructs the calculator to determine the tangent of the displayed angle
arc	Inverse trigonometric	Instructs the calculator to determine the angle of the selected trig function whose value is the displayed quantity, when pressed as a prefix to the sin, cos, or tan key
$\log x$	Common logarithm	Instructs the calculator to determine the logarithm to the base 10 of the displayed number
$\ln x$	Natural logarithm	Instructs the calculator to determine the logarithm to the base e of the displayed number
e^x	e to the x power	Instructs the calculator to raise the value of e to the displayed power
y^x	y to the x power	Instructs the calculator to raise y, the first entered number, to the power of x, the second entered number

Key Symbol	Key Name	Key Instruction
$x\sqrt{y}$	The xth root of y	Instructs the calculator to process y, the first entered number, to find the xth root. The value of x is the second entered number
STO	Store	Instructs the calculator to store the displayed number in the memory
RCL	Recall	Instructs the calculator to retrieve stored data from the memory
Σ	Sum and store	Instructs the calculator to algebraically add the displayed number to the number in the memory, and to store the sum in the memory
$x!$	Factorial	Instructs the calculator to find the factorial of the number displayed
$=$	Equals	(Algebraic entry method only) Instructs the calculator to complete the previously entered operation to provide the desired calculation result
CLx	Clear entry	Clears the last keyboard entry.
π	Pi	Enters the value of pi (π) in the display register
ENTER ↑	Enter	Loads contents of x register into y register and retains contents of x register in x register

1.6 THE PROGRAMMABLE POCKET CALCULATOR

The most sophisticated pocket calculators available at present are the
HP-55 and HP-65. They operate in Reverse-Polish notation, have memory
stacks and registers, uses floating-point arithmetic with scientific notation,
has an extensive three-level function set, and are programmable. From the
standpoint that the programmable pocket calculator implements logical
(Boolean) equations as well as Algebraic equations, can make logical
decisions, and will iteratively execute a preprogrammed set of instructions,
it can be correctly called a pocket computer. It is called a calculator only
because it does not satisfy the U.S. Government's import/export trade
definition of a computer. Because it is generally accepted that the definition of a computer (or calculator) changes as the state of the art of
computer design changes, it is also acknowledged that in 1955 the programmable pocket calculator would have been called a computer.

Programmable calculators provide a quantum jump in pocket computing
capability by making libraries of program listings and prerecorded magnetic tape programs available to the analyst at relatively low cost. These
libraries can be compiled by the user himself or can be purchased.

1.7 THE CALCULATORS TO BE DISCUSSED IN THIS BOOK

We have seen that there are three types of entry method, three types of
memory, and three kinds of number that can be implemented in any of the
three kinds of pocket calculator with (though not described here) four
types of function set and two types of $I/0$—the hard copy and the manual
$I/0$. Hence at least 432 types of calculator could be made up from
different combinations of these electronic hardware alternatives. While the
number of reasonable combinations is somewhat smaller, about 50, the
number of possible types of pocket calculator is still too large to be
covered in one book. We therefore analyze only three basic types of
hypothetical pocket calculators. One is a simple four-function calculator.
The other is an engineering machine, again a hypothetical one, but with a
function set characteristic of the SR 50/51 or HP-21/35/45 series
machines. The third is the programmable pocket calculator, which we
assume to have a four-register stack with a nine-register addressable
storage and a 100-word instruction set. This, too, is a hypothetical machine
whose properties are defined in the context of the discussion.

Of the three hypothetical pocket calculators, emphasis is placed on the second—the engineering-type four-register stack machine with the usual complement of engineering functions. Also, because the simple four-function machine is now available at very little cost, attention is given throughout the book to performing advanced analysis on this machine. What continues to amaze the writer is the extent to which the four-function pocket calculator can be applied to engineering analysis once the equations to be solved are manipulated in forms that require no memory for their evaluation.

For all these machines, we assume that we are limited at most to a 10-digit register and we use floating-point arithmetic with scientific notation.

The keyboard for the hypothetical four-function calculator to be discussed is sketched in Figure 1-1. The keyboard functions for the scientific and programmable pocket calculator are shown in Figure 1-2. The basis for the discussions dealing with this calculator is the HP-65, in that it is representative of what will be available in the foreseeable future.

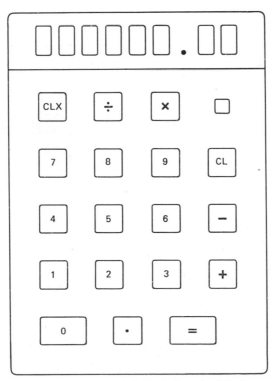

Figure 1-1 A hypothetical four-function pocket calculator keyboard.

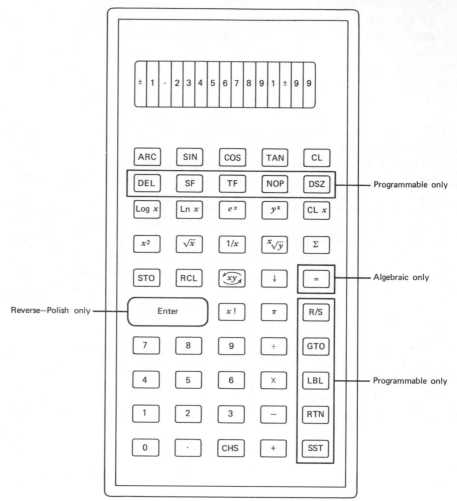

Figure 1-2 A hypothetical scientific pocket calculator keyboard (mixed Algebraic and Reverse-Polish and programmable functions).

The display details for all calculators discussed here are shown in Figure 1-3. The display features that we will discuss from time to time include the following:

Decimal point   Assumed to be to the right of any number entered unless positioned in another sequence with the ⬚ key.

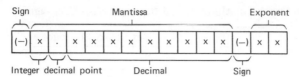

Figure 1-3 Typical pocket calculator display format.

Minus sign	Appears to the left of the 10-digit mantissa for negative numbers, and appears to the left of the exponent for negative exponents
Overflow indication	In most pocket calculators, the largest number that can be entered in the calculator is $\pm 9.999999999 \times 10^{99}$ without an overflow when a function is pressed. If a calculation result is larger than this value, the display will flash or give a numerical indication of overflow
Underflow indication	If a number closer to zero than to $\pm 1.0 \times 10^{-99}$ is entered in the calculator, the display will flash or indicate an underflow.

While concentrating on the hypthetical machines just mentioned, we shall comment on machines with slightly different keyboards where appropriate.

1.8 ARITHMETIC CALCULATIONS AND LANGUAGES

It might seem that the arithmetic functions of addition, subtraction, multiplication, and division are so basic to the pocket calculator that very little need be said about them. It is because they are so basic that they are discussed in some detail here. Arithmetic performed in one language is substantially different from that performed in another language. In one language certain arithmetic calculations are quite convenient and easy to remember to the infrequent user. Another language, though less convenient to the beginner, is more powerful and flexible (hence more convenient) to the frequent user. Finally, and perhaps most important, mixed arithmetic calculations illuminate the need for memory—whether manual, (using a scratch pad) or temporary data storage, (using automatic stacked registers) or permanent data storage, (using addressable memory).

Table 1-1 Arithmetic in Algebraic and Reverse-Polish Languages

Task	Key Stroke Sequence	
	Algebraic	Reverse-Polish
Sum $A \& B$	$A + B =$	$A \uparrow B +$
Sum $A \& B \& C$	$A + B + C =$	$\begin{cases} A \uparrow B + C + \\ A \uparrow B \uparrow C + + \end{cases}$
Sum $A \& B \& C \& D$	$A + B + C + D =$	$\begin{cases} A \uparrow B + C + D + \\ A \uparrow B \uparrow C + + D + \\ A \uparrow B \uparrow C + D + + \\ A \uparrow B \uparrow C \uparrow D + + + \end{cases}$
Multiply $A \& B$	$A \times B =$	$A \uparrow B \times$
Multiply $A \& B \& C$	$A \times B \times C =$	$\begin{cases} A \uparrow B \times C \times \\ A \uparrow B \uparrow C \times \times \end{cases}$
Multiply $A \& B \& C \& D$	$A \times B \times C \times D =$	$\begin{cases} A \uparrow B \times C \times D \times \\ A \uparrow B \uparrow C \times \times D \times \\ A \uparrow B \uparrow C \times D \times \times \\ A \uparrow B \uparrow C \uparrow D \times \times \times \end{cases}$
Compute $(A \times B) + (C \times D)$	$\begin{cases} A \times B \div D + C \times D \\ = \text{(no memory)} \\ A \times B + C \times D \\ = \text{(with hierarchy)} \\ A \times B \, \text{STO} \, C \times D \, \text{RCL} + \\ = \text{(with memory)} \end{cases}$	$\begin{cases} A \uparrow B \times C \uparrow D \times + \\ A \uparrow B \uparrow C \uparrow D \times \text{R} \downarrow \times \text{R} \uparrow + *^a \\ \cdot \\ \cdot \\ \cdot \end{cases}$
Compute $(A + B) \times (C + D)$	$A \times B \, \text{STO} \, C + D \times \text{RCL} =$	$\begin{cases} A \uparrow B + C \uparrow D + \times \\ A \uparrow B \uparrow C \uparrow D + \text{R} \downarrow + \text{R} \uparrow \times * \\ \cdot \\ \cdot \\ \cdot \end{cases}$

^{a}*See page 21 for a definition of R↓ and R↑.

The two most popular languages used in pocket calculators are algebraic and reverse-polish. The languages were introduced in Section 1.3. Table 1-1 illustrates the key strokes involved in performing additions, multiplications, and mixed arithmetic calculations such as products-of-sums and sums-of-products using both the algebraic language and the reverse-polish language. A number of insights on analysis on the various pocket calculators can be derived by examining the table. The most obvious is that the algebraic language programs the calculation of simple series arithmetic calculations in exactly the manner in which we would write them as an algebraic expression reduced to its simplest form. It is equally obvious that even simple series arithmetic calculations can be performed in a number of different ways when using reverse-polish language (except for the simplest operations of adding and multiplying two numbers). In a sense, then, for these simple arithmetic tasks, the algebraic language has one unique sequence of key strokes for performing the task, while the reverse-polish does not. When viewed from the algebraic language enthusiast's standpoint, this ambiguity in ways to solve simple series arithmetic problems in reverse-polish is viewed as a possible confusion factor for the pocket calculator user. The reverse-polish language enthusiast, however, views the same property as a measure of the flexibility of the reverse-polish notation. From his viewpoint, the user has greater flexibility in the algebraic forms in which an arithmetic problem can be presented for numerical evaluation. Furthermore, he could argue, the first form shown in each of the series calculations in Table 1-1 is close to the algebraic language key strokes, differing only in the second and last key strokes.

It is interesting that this distinction should come up at all, since the mixed arithmetic in the last two examples in Table 1-1 shows the many different ways in which the sum of products can be evaluated with the algebraic and reverse-polish languages. Note that the first example of the use of algebraic language to evaluate the sum of products illustrates the rewriting of the algebraic form as

$$(A \times B) + (C \times D) \equiv \left(\frac{A \times B}{D} + C \right) D$$

We see from the sequence of key strokes that the sum of products can be evaluated without memory. This form of evaluating the sum of products is ideal for use on the simple four-function calculators in that it requires no scratch pad memory and is within the set of operations available on even the simplest pocket calculator. A similar expression can be developed for calculating the product of sums without need of memory. Again, the algebraic form of the equation must be rewritten to be convenient for calculator evaluation as

$$(A+B)\times(C+D)=\left(\frac{(A+B)\times C}{D}+A+B\right)D$$

The importance of rewriting expressions in forms that are easily evaluated on the pocket calculator is obvious, however. The example of the sum of products (the second from the last in Table 1-1) shows that the most convenient form for implementation on any pocket calculator may depend on the language that that calculator uses and the sophistication with which it is implemented. For example, the second sequence of key strokes to evaluate the sum of products is in the standard algebraic form. This form works well for sums of products where the algebraic language is implemented with a hierarchy of operands, that is, the multiplies are performed before the sums. Also, the third example of the employment of algebraic language for evaluating the sum of products shows the standard algebraic forms for evaluating the sum of products on a machine that uses the algebraic language but has an additional register for memory.

We observe in Table 1-1 also that no memory is required for performing simple arithmetic calculations in algebraic language until we reach the product of sums, the last example in the table. Such is not the case for the reverse-polish language. For example, only two registers are required for implementing the simple sum $A+B+C+D$ in algebraic language. In reverse-polish language, only two registers are required to implement the sum as shown by the first example in the column of possible implementations of this series of sums. The other three possible implementations require additional registers in which to store the data A, B, C, and D. Clearly in algebraic language additional registers would not permit alternative ways to evaluate the sum, while in reverse-polish every additional register leads to one additional way. In the example shown in Table 1-1 it is assumed that there are four registers in which to store the four data A, B, C, and D. Obviously, the use of reverse-polish language with stacks of data registers adds flexibility to a pocket calculator. In a sense, then, polish notation and stacks go together in a pocket calculator. It is also apparent that algebraic language eliminates the need for extensive stacks of data registers, since no additional flexibility is permitted with the addition of register stacks. Therefore, most calculators that use algebraic language have smaller memories than pocket calculators using reverse-polish.

Another observation that we can make from Table 1-1 is that machines with algebraic notation which also have hierarchy and an additional register of memory (such as the SR-50) embody the highest level of capability available for pocket calculators using the algebraic language.

Such algebraic machines compete effectively in conducting mixed arithmetic calculations with the reverse-polish language machine, such as the HP35/45 series, with somewhat less electronic complexity. However, the reverse-polish with stacks adds operational flexibility for the user, which the algebraic machine does not. Moreover, the algebraic machine requires that the form of the equation be evaluated, particularly if it is highly complex. The reverse-polish language, on the other hand, provides the flexibility to evaluate very complex expressions with minimum attention being paid to the arrangement of terms. This flexibility is in part due to additional arithmetic registers that the typical reverse-polish machines generally have.

Because the manipulation of data among the data registers is essential to understanding both the reverse-polish with stacks machine and the advanced algebraic machines, we discuss memory manipulations next.

When we speak of a reverse-polish machine with stacks, we assume that a stack consists of four registers for storing numbers. Following Hewlett-Packard notation, we call these registers X, Y, Z, and T. Register X is at the bottom of the stack, T is at the top of the stack, and the display always shows the number in the X register. We designate the number in the register by the same letter in *italic* type. Thus X, Y, Z, and T are the contents of registers X, Y, Z, and T. When a number key is stroked, the number enters the X register which is displayed. The number is repeated in the Y register when the "enter" key $\boxed{\uparrow}$ is stroked. Whatever is in the Y register is "pushed up" into the Z register. The contents of the Z register are moved into the T register, and the contents of the T register are lost (see Figure 1-4). As data are entered into the Y register from the X register, the data in the other registers are "pushed up" automatically with the only data lost being the data in the T register. Data in the Y register can be viewed in the display by rolling the data from the Y register down to the X register by stroking the "roll-down" key $\boxed{R\downarrow}$. The data in the X register are then worked backwards in the stack to move to the top register (T), the data in the top register move to the Z register, the data in the Z register

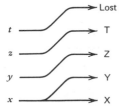

Figure 1-4 Data flow associated with data entry.

move to the Y register, and, as mentioned before, the data in the Y register move into the X register where they are displayed. Stroking the "roll-down" key again causes the data that were formerly in the Z register, which have been moved to the Y register, to move down to the X register where they can be seen in the display. All other data are moved to a neighboring register in the direction in which the roll is made. It follows that after four "roll-down" key strokes the stack will be arranged back in the original order where X is in its original location and is displayed in the X register, Y is in its original location, Z is in its original location, and T is in its original location. Stroking the "roll-down" key moves the data in the registers in the direction from the Y register to the X register. Stroking the "roll-up" key $\boxed{R\uparrow}$ moves the data in the direction from the X register to the Y register. The data flow associated with the data entry and "roll-down" and "roll-up" operations is seen in Figure 1-5.

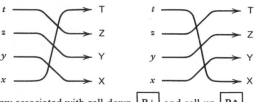

Figure 1-5   Data flow associated with roll-down $\boxed{R\downarrow}$ and roll-up $\boxed{R\uparrow}$.

Another commonly used stack manipulation is the replacement of the data in the X register with the data in the Y register and vice versa. The data flow associated with stroking the "X-Y exchange" key $\boxed{\circlearrowright}$ is sketched in Figure 1-6.

The data flow associated with the stack operations, when performing

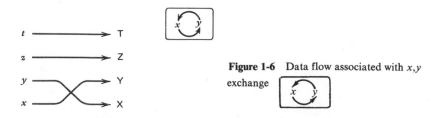

Figure 1-6   Data flow associated with x,y exchange

addition, subtraction, multiplication, and division, is sketched in Figure 1-7. We see the following:

1. For summation, the contents of the Y and X registers are added and displayed in the X register.
2. For subtraction, the contents of the X register are subtracted from the contents of the Y register and displayed in the X register;
3. For multiplication, the contents of the X register are multiplied by the contents of the Y register and displayed in the X register; and
4. For division, the contents of the Y register are divided by the contents of the X register and displayed in the X register.

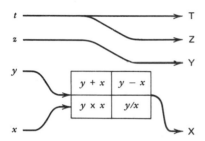

Figure 1-7 Data flow associated with $+$, $-$, $\times$ and $\div$.

For these basic four functions the contents of the T register are always retained and never lost. This feature of the operational stack is very useful for certain repeated calculations.

It is worth pointing out here that many of the functions evaluated by a single key stroke on the typical Reverse-Polish with stacks pocket calculators result in the loss of some data in the operational stack. For instance, in the HP35/45 series calculators, the contents of the T registers are lost when evaluating trigonometric functions. They are retained when evaluating logarithmic and algebraic functions, such as taking the square root, taking the inverse, taking the logarithms, or exponentiating.

Figures 1-8 and 1-9 illustrate the typical data flow in the stacks when the product of two sums and the sum of two products are evaluated using reverse-polish with stacks. Figure 1-8a shows the usual procedure for evaluating the sum of products, which does not involve the use of the top register. To illustrate the flexibility of the reverse-polish with stacks and operations associated with the top register, Figure 1-8b shows the same calculations using the "roll-up" and "roll-down" features of the stack manipulations. Figures 1-10 and 1-11 present the typical data flow

Figure 1-8a

Register	1	2	3	4	5	6	7	8	9
T									
Z						(A×B)	(A×B)		
Y		A	A		(A×B)	C	C	(A×B)	
X (Display register)	A	A	B	(A×B)	C	C	D	(C×D)	(A×B)+(C×D)
Key	A	↑	B	×	C	↑	D	×	+
Step	1	2	3	4	5	6	7	8	9

Figure 1-8a Data flow associated with the sum of two products $(A \times B)+(C \times D)$ using key strokes $A \uparrow B \times C \uparrow D \times +$ on a Reverse-Polish Machine.

Figure 1-8b

Register	1	2	3	4	5	6	7	8	9	10	11	12
T						A	A	A	(C×D)	(C×D)	A	A
Z				A	A	B	B	A	A	A		
Y		A	A	B	B	C	C	B	A		(A×B)	
X (Display register)	A	A	B	B	C	C	D	(C×D)	B	(A×B)	(C×D)	(A×B)+(C×D)
Key	A	↑	B	↑	C	↑	D	×	R↓	×	R↑	+
Step	1	2	3	4	5	6	7	8	9	10	11	12

Figure 1-8b Data flow associated with the sum of two products $(A \times B)+(C \times D)$ using key strokes $A \uparrow B \uparrow C \uparrow D \times R \downarrow \times R \uparrow +$.

	1	2	3	4	5	6	7	8	9
T									
Z						$(A+B)$	$(A+B)$		
Y		A	A		$(A+B)$	C	C	$(A+B)$	
X	A	A	B	$(A+B)$	C	C	D	$(C+D)$	$(A+B)\times(C+D)$
Key	A	$\uparrow$	B	$+$	C	$\uparrow$	D	$+$	$\times$
Step	1	2	3	4	5	6	7	8	9

Figure 1-9 Data flow associated with $(A+B)\times(C+D)$ using key strokes $A\uparrow B + C\uparrow D + \times$ on a Reverse-Polish machine.

25

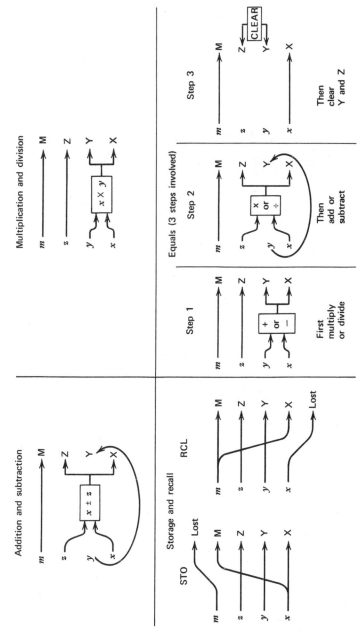

Figure 1-10 Data flow associated with keyboard functions of an Algebraic machine with memory and hierarchy ("multiply before add").

26

	Step 1	Step 2	Step 3	Step 4	Step 5	Step 6	Step 7	Step 8	Step 9	Step 10	Step 11	Step 12
M					$(A+B)$	$(A+B)$	$(A+B)$	$(A+B)$	$(A+B)$	$(A+B)$		$(A+B)$
Z		A	A				C	C				
Y										$(C+D)$	$(C+D)$	
X	A	A	B	$(A+B)$	$(A+B)$	C	C	D	$(C+D)$	$(C+D)$	$(A+B)$	$(A+B)\times(C+D)$
Key	A	$+$	B	$=$	STO	C	C	$+$	D	$=$	$\times$	RCL
Step	1	2	3	4	5	6	7	8	9	10	11	12

Figure 1-11 Data flow associated with $(A+B)\times(C+D)$ using key strokes $A+B=\text{STO}\,C+D=\times\text{RCL}=$ on an algebraic machine with memory and hierarchy ("multiply before add").

27

associated with keyboard functions and a calculation of the product of two sums using algebraic with memory. A comparison of Figures 1-8, 1-9, 1-10, and 1-11 indicates clearly that the greater the memory storage capacity in a pocket calculator the greater the flexibility in its use.

The question of languages in pocket calculators is akin to that in minicomputers or large computers, or different nationalities for that matter —the language you know the best is the language you like the most, unless you have sufficient multilingual skills to recognize the subtle advantages of one language over another. What matters least is the type of language or size of memory associated with any specific pocket calculator; what matters most is to begin to use some pocket calculator in advanced analysis. The solid-state revolution has enabled the engineer to perform fairly sophisticated analysis at his desk, in his home, or on a trip, without the need for access to a computing facility. Simply stated, those who capitalize on this aspect of the solid-state revolution and keep current with the development of pocket computing machines will have a tremendous advantage over those who do not.

1.9 SCIENTIFIC KEYBOARD FUNCTION EVALUATION

In this section we use the four-function calculator to evaluate the scientific functions normally found on the scientific calculator keyboard. The sine, cosine, tangent, exponential, logarithmic, arc sine, arc cosine, and arc tangent functions are presented in nested parenthetical forms in two different ways. The first way is in the nested parenthetical form of the truncated series approximations of these functions. The second way is in a curve-fit polynomial form that permits precision evaluation of these functions over a broader range than the simple series expansions of the functions. Also covered here are the algebraic functions of raising a number to its nth power or evaluating its nth root.

Raising a Number to a Power

Raising a number to a power on a four-function calculator can be done simply by repeated multiplication. A fairly high power, such as 100, involves 100 data entries and 100 multiplies, which result in many key strokes and many possibilities for error. An alternative is to use the constant key available on many of the four-function calculators. The constant key is built into these calculators to make it convenient to multiply or divide a series of numbers by a constant number. In the case of raising a number to a power, we put the chain-constant switch to the constant position and then input the number to the X register, depress the constant key, and then raise the number to the power n by stroking the equals key n times. This approach virtually eliminates the error associated with repeated data entry in the primitive n-multiply approach. Even in this case, however, raising a number to the power of 100 involves 100 depressions of the equals key (an error prone procedure). This can be circumvented by breaking down the power into its prime factors and performing nested parenthetical multiplies to evaluate numbers raised to high powers. For example, suppose that we wish to raise π to the 100th power with only a single entry of π. This can be conveniently done on the simple four-function calculator by noting that the prime factors of 100 are 2, 2, 5, 5. Then

$$(\pi)^{100} = \left(\left(\left((\pi)^2\right)^2\right)^5\right)^5$$

$$= \left(\left((9.86960440)^2\right)^5\right)^5$$

$$= \left((97.40909108)^5\right)^5$$

$$= (8.769956822 \times 10^9)^5$$

$$= 5.187848391 \times 10^{49}$$

On pocket calculators that have the squaring operator, numbers can be raised to any integer power through the 10th by entering the data only three times. Since the prime factors of many of our exponents are made up of 2, 3, 5, and 7, raising numbers to these powers involves only two data entries.

Computing roots on the four-function calculator requires iterative operations. Among the various approaches to evaluating roots, the simplest is Newton's method. Though this method leaves much to be desired in most applications, it can be used conveniently for computing the roots of numbers. We have more to say on this topic later in the book. For now, note that the formula for computing the roots is

$$x_{k+1} = \frac{1}{n}\left(x_k\left(\frac{N}{x_k^n} + n - 1\right)\right)$$

where $x_k = k$th estimate of $\sqrt[n]{N}$
This equation requires an initial approximation, which is used to develop a

second, more exact, approximation, which in turn is again used to develop a third, even more exact, approximation. The process usually converges quickly when the initial estimate of the nth root is known. Convergence can be markedly slow, however, when the first estimate is not fairly close to the root in question. Examples of the convergence properties of the use of Newton's method for evaluating the third, fifth, and seventh roots of π are shown in Table 1-2.

Table 1-2   Examples of the Convergence of Newton's Method for nth Roots of π

Number of Iterations	$\sqrt[3]{\pi}$	$\sqrt[5]{\pi}$	$\sqrt[7]{\pi}$
0	1.0	1.0	1.0
1	1.713864218	1.428318531	1.305941808
2	1.499089493	1.293620977	1.209849586
3	1.465379670	1.259260005	1.180121812
4	1.464592311	1.257280369	1.177679333
5	1.464591888	1.257274116	1.177664031
6	1.464591888	1.257274116	1.177664030
7	—	—	1.177664030
Check by computing π	3.141592656	3.141592658	3.141592655
π (actual)	3.141592654	3.141592654	3.141592654
Absolute error	0.000000002	0.000000004	0.000000001

Close examination of Table 1-2 shows that for an initial guess of 1, the process converges in five iterations to the accuracy of the pocket calculator. However, a determination of this requires a sixth iteration, and in the case of the seventh root also a seventh iteration. The table also shows the check of the root by repeated multiplies and the comparison with the true value of π, indicating an accuracy of 1 part in 10^8 after only six or seven iterations. In general, this method cannot be expected to converge so quickly for other functions. It happens to converge quickly for the nth root function because of the nice properties of that function. Note that Newton's formula for computing the nth root also works for the simple square root. This can be seen by setting n equal to 2 in the equation. Then

Newton's formula for computing the square root iteratively (which is also due to Joseph Raphson, a contemporary of Newton's—hence this method is often called the Newton-Raphson technique) gives the equation

$$(\sqrt{N})_{k+1} = \frac{1}{2}\left(\frac{N}{(\sqrt{N})_k} + (\sqrt{N})_k\right)$$

which can be used for iteratively computing the square root of a number N.

Nested Parenthetical Forms

Many functions of interest to engineers can be written in a power series. This series can be generated by using Taylor's theorem, Maclaurin's theorem, Chebyshev polynomials, and so on. Furthermore, an empirical data set can be fit with power series. When so written, they take the "standard" form

$$f(x) = a_0 + a_1 x + a_2 x^2 + a_3 x^3 + \cdots + a_n x^n + \cdots \qquad (1\text{-}1)$$

If we were to evaluate this series in the most straightforward manner on the simple four-function calculator, we would compute each term in the series and record it on a scratch pad. When all the terms of interest were evaluated, the sum would be computed on the pocket calculator. The number of key strokes involved for a 10-digit data entry is shown in Table 1-3. The total number of key strokes for data entry plus instruction are

$$\text{Total key strokes} = \sum_{i=1}^{n} 12i + 10 = 6n^2 + 16n \qquad (1\text{-}2)$$

Table 1-3   Key Strokes Required to Evaluate Power Series in Standard Form

Operation	Key Strokes
Record,[a] a_0	0
Compute and record, $a_1 x$	22
Compute and record, $a_2 x^2$	34
Compute and record, $a_3 x^3$	46
$\vdots$	$\vdots$
Compute and record, $a_n x^n$	$12n + 10$

[a]No one would input a_0 in the calculator and then recopy it on a scratch pad.

assuming that each data entry involved the full register. Clearly for $n > 3$ the number of key strokes becomes laboriously large and for $n > 5$ the chances for error become enormous. By rewriting equation 1-1 in the form

$$a_0 + x\left(a_1 + x\left(a_2 + x\left(a_3 + \cdots + x\left(a_{n-2} + x\left(a_{n-1} + a_n x\right)\right) \cdots \right)\right)\right) \quad (1\text{-}3)$$

we reduce the number of key strokes, because the formula is organized in the natural language of the machine and requires no scratch pad storage. In pocket computer instructions, equation 1-3 would be evaluated working from the inside out with the following instruction set:

$$a_n \;\boxed{\times}\; x \;\boxed{+}\; a_{n-1} \;\boxed{\times}\; x \;\boxed{+}\; a_{n-2} \;\boxed{\times}\; x \;\boxed{+}\; \cdots$$

$$\boxed{+}\; a_2 \;\boxed{\times}\; x \;\boxed{+}\; a_1 \;\boxed{\times}\; x \;\boxed{+}\; a_0 \;\boxed{=}$$

The number of key strokes for this evaluation of equation 1-1 is

Number of terms	1*	2	3	$\cdots$	n
Total key strokes	0	33	55	$\cdots$	$10(2n-1) + 2n - 1$

And the key strokes for data entry plus instructions total

$$\text{total key strokes} = 11(2n - 1) \qquad (1\text{-}4)$$

We see, then, that in nested parenthetical form we can carry up to 10 terms before the dimensions of the problem get out of hand—that is, up to six more terms than in the "standard" form. This business with the forms of equations is worth remembering for series evaluation on any calculator or computer in that the nested parenthetical forms are generally processed *faster* than are standard forms when computing time is involved. This is because the number of arithmetic operations grows as the square of the number of terms for series written in standard form and only proportional to the number of terms for series written in nested parenthetical forms.

Note also that it is unnecessary to use a scratch pad when evaluating series in parenthetical forms, since the operands and operations are in the

*No one would evaluate a single-term series on the calculator.

appropriate order for evaluation with algebraic, polish, or reverse-polish entry methods.

Comparing equations 1-2 and 1-3 we see that the nested parenthetical form substantially reduces the number of key strokes by reducing the number of data entries required for the calculation. Even more dramatic is the impact that rewriting the equation in nested parenthetical form has on the time required to perform the numerical evaluation of the power series on a pocket calculator. If we assume that, on the average, for every key stroke and digit record the calculation takes 1 second, we would expect the nested parenthetical form to involve $22/(6n+32)\%$ (for $n>5$) of the time required for a standard-form power series evaluation. In general, nested parenthetical forms of power series or polynomials are more quickly evaluated than the standard forms. The more common scientific keyboard functions can be evaluated on the four-function calculator using the following nested formulas:

$$\ln(1+x) \cong x\left(1 - \frac{x}{2}\left(1 - \frac{2x}{3}\left(1 - \frac{3x}{4}\left(1 - \frac{4x}{5}\right)\right)\right)\right), \qquad (|x|<1)$$

$$\ln(x) \cong y\left(1 + \frac{y}{2}\left(1 + \frac{2y}{3}\left(1 + \frac{3y}{4}\left(1 + \frac{4y}{5}\right)\right)\right)\right), \qquad y = \left(\frac{x-1}{x}\right), \quad (x > \tfrac{1}{2})$$

$$\ln(x) \cong y\left(1 - \frac{y}{2}\left(1 - \frac{2y}{3}\left(1 - \frac{3y}{4}\right)\right)\right), \qquad y = (x-1), \qquad (|x-1| \leq 1)$$

$$\ln\left(\frac{x+1}{x-1}\right) \cong \frac{2}{x}\left(1 + \frac{1}{6x^2}\left(1 + \frac{3}{5x^2}\left(1 + \frac{5}{7x^2}\left(1 + \frac{7}{9x^2}\right)\right)\right)\right), \qquad |x| \geq 1$$

$$e^x \cong 1 + x\left(1 + \frac{x}{2}\left(1 + \frac{2x}{6}\left(1 + \frac{6x}{24}\left(1 + \frac{24x}{120}\left(1 + \frac{120}{720}x\right)\right)\right)\right)\right)$$

$$\sin(x) \cong x\left(1 - \frac{x^2}{6}\left(1 - \frac{6x^2}{120}\left(1 - \frac{120x^2}{5040}\left(1 - \frac{5040}{362880}x^2\right)\right)\right)\right)$$

$$\cos(x) \cong \left(1 - \frac{x^2}{2}\left(1 - \frac{2x^2}{24}\left(1 - \frac{24x^2}{720}\left(1 - \frac{720}{40320}x^2\right)\right)\right)\right)$$

$$\tan(x) \cong x\left(1 + \frac{x^2}{3}\left(1 + \frac{6x^2}{15}\left(1 + \frac{255}{630}x^2\right)\right)\right)$$

$$\cotan(x) \cong \frac{1}{x} - \frac{x}{3}\left(1 + \frac{3x^2}{45}\left(1 + \frac{90}{945}x^2\right)\right)$$

$$\arcsin(x) \cong x\left(1 + \frac{x^2}{6}\left(1 + \frac{18x^2}{40}\left(1 + \frac{600}{10008}x^2\right)\right)\right), \qquad |x| \leqslant 1$$

$$\arctan(x) \cong x\left(1 - \frac{x^2}{3}\left(1 - \frac{3x^2}{5}\left(1 - \frac{5x^2}{7}\right)\right)\right), \qquad x^2 < 1$$

$$\arctan(x) \cong \frac{\pi}{2} - \frac{1}{x}\left(1 - \frac{1}{3x^2}\left(1 - \frac{3}{5x^2}\right)\right), \qquad |x| > 1$$

These formulas were selected on the basis of the reasonableness of their intervals of convergence. The four different approximations for the natural logarithm span the region from $x = -1$ to $+\infty$. They are all written in convenient nested parenthetical forms and can be used for immediate evaluation on the pocket calculator. This table, if copied and reduced, can be conveniently taped to the back of your pocket calculator for handy reference.

Another approach to evaluating these scientific functions is to use a curve-fit polynomial over a broad range of the argument. Such polynomials are tabulated in Table 1-4 for the functions on the keyboard of the scientific pocket calculator. These polynomials will permit precise evaluation of the logarithmic, exponential, and trigonometric functions on the four-function calculator and thus make it capable of performing any analysis that can be performed on the scientific pocket calculator.

To put functions into forms that are easily computed on the pocket calculator, use the following procedures:

Procedure 1

    (a)   Either find or generate a table of values for the function of interest to the accuracy of interest.

    (b)   Prepare an interpolating polynomial (see Chapter 2) that passes through selected points of interest in the table but spans the range of interest in the argument.

    (c)   Identify the maximum error of the polynomial approximation on the interval of interest.

    (d)   If the accuracy is satisfactory, write the polynomial in nested parenthetical form, and then use it for approximate evaluation of the function on the pocket calculator.

Table 1-4 Polynomial Approximations of Many Functions Found on the Keyboard of the Scientific Pocket Calculators

---

(1) $\text{Log}_{10}(x) = t(a_1 + t^2(a_3 + t^2(a_5 + t^2(a_7 + a_9^2)))) + \epsilon(x)$

Here

$$t = (x-1)(x+1)^{-1}$$

and

$$|\epsilon(x)| \leqslant 10^{-7} \quad \text{where} \quad 10^{-1/2} \leqslant x \leqslant 10^{+1/2}$$

for

$$a_1 = 0.868591718 \qquad a_7 = 0.094376476$$
$$a_3 = 0.289335524 \qquad a_9 = 0.191337714$$
$$a_5 = 0.177522071$$

(2) $\text{Log}_{10}(x) = t(a_1 + a_3 t^2) + \epsilon(x)$
where $t = (x-1)(x+1)^{-1}$, $a_1 = 0.86304$, and $a_3 = 0.36415$.
Then

$$|\epsilon(x)| \leqslant 6 \times 10^{-4} \quad \text{where} \quad 10^{-1/2} \leqslant x \leqslant 10^{+1/2}$$

(3) $\text{Ln}(1+x) = x(a_1 + x(a_2 + x(a_3 + x(a_4 + a_5 x)))) = \epsilon(x)$

Here

$$a_1 = \quad 0.99949556 \qquad a_4 = -0.13606275$$
$$a_2 = -0.49190896 \qquad a_5 = \quad 0.03215845$$
$$a_3 = \quad 0.28947478$$

Then

$$|\epsilon(x)| \leqslant 10^{-5} \quad \text{where} \quad 0 \leqslant x \leqslant 1$$

(4) $\text{Ln}(1+x) = x(a_1 + x(a_2 + x(a_3 + x(a_4 + x(a_5 + x(a_6 + x(a_7 + a_8 x))))))) + \epsilon(x)$

Here

$$a_1 = \quad 0.9999964239 \qquad a_5 = \quad 0.1676540711$$
$$a_2 = -0.4998741238 \qquad a_6 = -0.0953293897$$
$$a_3 = \quad 0.3317990258 \qquad a_7 = \quad 0.0360884937$$
$$a_4 = \quad 0.2407338084 \qquad a_8 = \quad 0.0064535442$$

and

$$|\epsilon(x)| \leqslant 3 \times 10^{-8} \quad \text{where} \quad 0 \leqslant x \leqslant 1$$

(5) $e^{-x} = 1 + x(a_1 + a_2 x) + \epsilon(x)$
where

$$a_1 = -0.9664 \quad \text{and} \quad a_2 = 0.3536$$

Table 1-4 *(Continued)*

Then

$$|\epsilon(x)| \leqslant 3 \times 10^{-3} \quad \text{where} \quad 0 \leqslant x \leqslant \ln 2$$

(6) $e^{-x} = 1 + x(a_1 + x(a_2 + x(a_3 + a_4 x)))$

where

$$a_1 = -0.9998684 \qquad a_3 = -0.1595332$$
$$a_2 = 0.4982926 \qquad a_4 = 0.0293641$$

Then

$$|\epsilon(x)| \leqslant 3 \times 10^{-5} \quad \text{where} \quad 0 \leqslant x \leqslant \ln 2$$

(7) $\text{Sin}(x) = x(1 + x^2(a_2 + a_4 x^2)) + x\epsilon(x)$

where

$$a_2 = -0.16605 \quad \text{and} \quad a_4 = 0.00761$$

Then

$$|\epsilon(x)| \leqslant 2 \times 10^{-4} \quad \text{where} \quad 0 \leqslant x \leqslant \frac{\pi}{2}$$

(8) $\text{Sin}(x) = x(1 + x^2(a_2 + x^2(a_4 + x^2(a_6 + x^2(a_8 + a_{10} x^2))))) + x\epsilon(x)$

where

$$a_2 = -0.1666666664 \qquad a_8 = 0.0000027526$$
$$a_4 = 0.0083333315 \qquad a_{10} = -0.0000000239$$
$$a_6 = -0.0001984090$$

Then

$$|\epsilon(x)| \leqslant 2 \times 10^{-9} \quad \text{where} \quad 0 \leqslant x \leqslant \frac{\pi}{2}$$

(9) $\text{Cos}(x) = 1 + x^2(a_2 + a_4 x^2) + \epsilon(x)$

where

$$a_2 = -0.49670$$
$$a_4 = 0.03705$$

Then

$$|\epsilon(x)| \leqslant 9 \times 10^{-4} \quad \text{where} \quad 0 \leqslant x \leqslant \frac{\pi}{2}$$

(10) $\text{Cos}(x) = 1 + x^2(a_2 + x^2(a_4 + x^2(a_6 + x^2(a_8 + a_{10} x^2)))) + \epsilon(x)$

where

$$a_2 = -0.4999999963 \qquad a_8 = 0.0000247609$$
$$a_4 = 0.0416666418 \qquad a_{10} = -0.000002605$$
$$a_6 = -0.0013888397$$

Table 1-4 *(Continued)*

Then
$$|\epsilon(x)| < 2 \times 10^{-9} \quad \text{where} \quad 0 < x < \frac{\pi}{2}$$

(11) $\text{Tan}(x) = x(1 + x^2(a_2 + a_4 x^2)) + x\epsilon(x)$
where

$$a_2 = 0.31755$$
$$a_4 = 0.20330$$

Then

$$|\epsilon(x)| < 10^{-3} \quad \text{where} \quad 0 < x < \frac{\pi}{4}$$

(12) $\text{Tan}(x) = x(1 + x^2(a_2 + x^2(a_4 + x^2(a_6 + x^2(a_8 + x^2(a_{10} + a_{12}x^2)))))) + x\epsilon(x)$
where

$a_2 = 0.3333314036$	$a_8 = 0.0245650893$
$a_4 = 0.1333923995$	$a_{10} = 0.0029005250$
$a_6 = 0.0533740603$	$a_{12} = 0.0095168091$

Then
$$|\epsilon(x)| < 2 \times 10^{-3} \quad \text{where} \quad 0 < x < \frac{\pi}{4}$$

(13) $\text{Cotan}(x) = \dfrac{1}{x}(1 + x^2(a_2 + a_4 x^2)) + \dfrac{\epsilon(x)}{x}$

where

$$a_2 = -0.332867$$
$$a_4 = -0.024369$$

Then

$$|\epsilon(x)| \leqslant 3 \times 10^{-5} \quad \text{where} \quad 0 \leqslant x \leqslant \frac{\pi}{4}$$

(14) $\text{Cotan}(x) = \dfrac{1}{x}(1 + x^2(a_2 + x^2(a_4 + x^2(a_6 + x^2(a_8 + a_{10}x^2))))) + \dfrac{\epsilon(x)}{x}$

where

$a_2 = -0.3333333410$	$a_8 = -0.0002078504$
$a_4 = -0.0222220287$	$a_{10} = -0.0000262619$
$a_6 = 0.0021177168$	

Then

$$|\epsilon(x)| \leqslant 4 \times 10^{-10} \quad \text{where} \quad 0 \leqslant x \leqslant \frac{\pi}{4}$$

Table 1-4 *(Continued)*

---

(15) $\mathrm{Arcsin}(x) = \dfrac{\pi}{2} - (1-x)^{1/2}(a_0 + x(a_1 + x(a_2 + a_3 x))) + \epsilon(x)$

where

$$a_0 = 1.5707288 \qquad a_2 = 0.0742610$$
$$a_1 = -0.2121144 \qquad a_3 = -0.0187293$$

Then

$$|\epsilon(x)| \leqslant 5 \times 10^{-5} \quad \text{where} \quad 0 \leqslant x \leqslant 1$$

(16) $\mathrm{Arctan}(x) = x(a_1 + x^2(a_3 + x^2(a_5 + x^2(a_7 + a_n x^2)))) + \epsilon(x)$

where

$$a_1 = 0.9998660 \qquad a_7 = -0.0851330$$
$$a_3 = -0.3302995 \qquad a_9 = 0.0208351$$
$$a_5 = 0.1801410$$

Then

$$|\epsilon(x)| \leqslant 10^{-5} \quad \text{where} \quad -1 \leqslant x \leqslant 1$$

If the tables are not available and there is not sufficient time to prepare them, use Procedure 2.

Procedure 2

(a) Prepare a series approximation of the function centered on the interval of interest.

(b) Using a Chebyshev polynomial economization scheme (see Chapter 8) reduce the order of the polynomial.

(c) Test the polynomial for accuracy over the argument's interval of interest.

(d) If the polynomial is not sufficiently accurate, include more terms in the original approximating polynomial before Chebyshev economization, then use the Chebyshev procedure and test the polynomial again.

(e) When the polynomial is sufficiently accurate, write it in nested parenthetical form and use it to evaluate the function on the pocket calculator.

The numerical methods associated with generating interpolating polynomials are discussed in Chapter 2. The Chebyshev economization procedure and approximation with rational polynomials are discussed in Chapter 8.

An interesting aside is that the logarithmic, exponential, and transcendental functions and their inverses and hyperbolic counterparts are typically generated in pocket calculators with pre-programmed, recursion algorithms. These algorithms generate the numerical values of these functions using CORDIC techniques*. A CORDIC technique does not implement series expansion approximating polynomials. They are hardware algorithms that generate the numerical values of the mathematical functions in which we are interested. In a word, function evaluation on the pocket calculator is done to high precision using computing techniques and algorithms that are convenient and efficient from a circuit implementation viewpoint more than an analytical viewpoint.

1-10 ACCURACY IN FUNCTION EVALUATION

Books on numerical analysis or computer calculations usually present the equations for propagating relative or absolute error through an analysis. In this book we take a slightly different approach. Our concerns here are working within the limitations of the pocket calculator's computing capability and understanding the calculator's impact on the generation of error that gets introduced into the problem. We wish to identify methods and techniques for getting around these problems.

The floating-point number system affects the calculations on the pocket calculator through its treatment of overflow and underflow. When a number exceeds the largest number in the calculator, the calculator is usually set to its largest number and the calculation is set to overflow the contents of the calculator. Similarly, when the calculation calls for a number that is smaller than the smallest number in the calculator, the number usually is set equal to zero and the calculation is set to underflow the machine's capability. Intuitively, replacing an underflow by zero seems more reasonable than replacing an overflow by the maximum number available in the calculator. However, one must be careful in such generalizations. Computing e^{228} using the inverse of e^{-228} is not the same as evaluating e^{228} directly. The reason is that e^{-228} is set equal to zero and thus the inverse is undefined, while e^{228} is within the number system of the pocket calculator.

$$\frac{1}{e^{-228}} = \frac{1}{0_{\text{underflow}}} \rightarrow \text{undefined}$$

$$e^{228} = 1.045061560 \times 10^{99}$$

---

*The Cordic trigonometric computing technique—IRE Transactions on Electronic Computer, September 1959.

What is surprising is that these number system "end effects" can lead to some practical limitations on the range of variables for which the function can be evaluated. Table 1-5 shows the effect of overflow and underflow on the range of the function $x^5 e^x/(e^x - 1)$.

Table 1-5   The Effect of Overflow and Underflow on the Range of Function Evaluation

x	$x^5 e^x / e^x - 1$	$x^5 / 1 - e^{-x}$
1	1.581976707	1.581976707
10	1.000045407 × 10^5	1.000045407 × 10^5
100	1.0 × 10^{10}	1 × 10^{10}
200	3.200000023 × 10^{11}	3.200000023 × 10^{11}
202	3.363232171 × 10^{11}	3.363232170 × 10^{11}
203	3.447308829 × 10^{11}	3.447308829 × 10^{11}
204	Overflow	2.533058573 × 10^{11}
220	Overflow	5.153631990 × 10^{11}
225	Overflow	5.766503900 × 10^{11}
226	Overflow	5.895792594 × 10^{11}
227	Overflow	6.027389914 × 10^{11}
228	Overflow	Underflow

The function is written in two ways in the table: favoring underflow and favoring overflow. That is, the function in the first column will eventually overflow the calculator's field of numbers because of the evaluation of $x^5 e^x$, while the function in the second column will eventually underflow the calculator's field of numbers because of the evaluation of e^{-x}. The table shows that the range of the variable x for which the function can be evaluated is limited sooner by the overflow effect than the underflow effect. In fact, the function written in the form that will eventually result in underflow can explore the range of the argument which is 12% greater than the same function that will eventually result in overflow. In general, pocket calculator analysis favors functions written in the form that will eventually underflow.

Roundoff Error

Roundoff error is similar to the end effects associated with underflow and overflow. While many understand roundoff, its practical impact on engineering-type calculations is often ignored with occasionally surprising results. Because some of the modern pocket calculators display mantissas to 13 places, it is easy to overlook the roundoff effect in a calculation, thinking that the calculator's large mantissa will certainly maintain accuracy through a sequence of calculations. The question here, then, is not how to round off a calculation but, rather, how does the roundoff introduce error in a practical manner in a calculation? Roundoff is an end effect. It is similar to underflow and overflow in that the last digit in the mantissa is arbitrarily changed to another number on the basis of some rationale. It is different from underflow and overflow effects: end effects associated with the number system in the calculator impact the range of the argument that can be examined; roundoff does not. Roundoff can actually propagate error into the most significant digits of the calculation. One might ask, "How does the roundoff of a three or four significant digit number propagate into the most significant digit?" This is precisely what we shall discuss here; an example of how roundoff in the third significant digit propagates to the first significant digit resulting in a 100% error is used to illustrate the problem.

Table 1-6 shows the calculation of the difference between products of numbers known accurately to three significant digits.

Table 1-6   Error from the Least Significant Digits to the most Significant Digits

Desired Calculation	Calculator Results	Rounded Calculator Results	Calculator Results Rounded
0.234×0.567	0.132 678	0.133	0.132 678
-0.232×0.567	-0.131 312	-0.131	-0.131 312
$0.xxx$	1.366×10^{-3}	2×10^{-3}	1×10^{-3}

Column 1 shows the desired calculation. Column 2 shows the results achieved on a pocket calculator, and column 3 shows the results achieved by first rounding each of the numbers generated in the product and then taking the difference. Column 4 shows first taking the difference between the unrounded numbers and then performing the rounding operation. Precisely what we mean here by rounding is the following. When the two three-digit numbers are multiplied, their product has either five or six places. Because the original numbers are only known to three places, we must drop two or three digits from the product. The rounding operation is adding 1 in the third place if the fourth-place digit is five or greater, or adding zero to the third digit if the fourth digit is less than five.

Now let us examine Table 1-6 closely. The desired calculation involves roundoff because the numbers in the products are only known to three places. The calculator results that are displayed to five or six places are really only known to three places and thus the number must be rounded. The result of taking the difference of the unrounded numbers is 1.366×10^{-3} which is only accurate to the first digit. If the second, third, and fourth digits are retained in additional calculations, they introduce a multiplication error into the problem that is propagated forward in any calculations. Clearly, the propagation of this type of error in an extended calculation can provide meaningless results. This roundoff error is well known and is not commonly made by most analysts.

It is the errors associated with the third-column calculations that are occasionally introduced into calculations. They arise from what seem to be reasonable calculations but are in fact mathematically incorrect and thus introduce substantial errors. The results in the first row of calculations are rounded to the third significant digit before the subtraction is performed, giving 2×10^{-3}. Column 4 shows the subtraction being performed before the roundoff is performed. It is apparent that the difference in the two calculations is a factor of 2 (100% difference in the two numbers). The rationale for the calculation of column 3 is that we really only know the number to three significant digits, and thus should round each product before subtracting. The rationale for the calculation in column 4 is that rounding arbitrarily changes one of the numbers in the calculation, which introduces an end-effect error. In column 3 there are two end-effect errors which can combine into a sizable resultant error, while in column 4 only one end-effect error occurs when the products and subtractions are completed and the result is rounded. In this sense, then, if a column of n products are taken followed by n subtractions, where roundoff is performed after the multiplication, there are n opportunities for propagating the roundoff effect from the third significant digit to the first significant digit. However, if the roundoff is performed after the subtractions are made, there is only one opportunity for propagating this roundoff error forward into the most significant digits. Thus the rule of thumb for accurate calculations is to roundoff on the last step. The example chosen here carries the roundoff error immediately from the last significant digit to the first significant digit, which usually is not the situation. It is worth pointing out, however, that calculations to one part in 10,000 involving differences can move roundoff error as much as three significant digits forward, thus modifying sensitivity analysis (evaluation of derivatives with finite differences) results in the third and even second places.

In summary, roundoff becomes a problem in calculations mainly when two numbers of the same size are subtracted. The roundoff propagates forward as a result of the cancellation of the leading digit in a subtraction process. This brings the roundoff errors from the least significant digit into the most significant digits. Because the display in the calculator's display window shows a mantissa to 13 places, the inexperienced analyst can be "spoofed" into assuming incorrectly that he has an accurate number.

Unfortunately, there is no systematic approach to analyzing the effect of roundoff in extensive calculations. All that can be said is that care must be taken not to write equations in forms leading to differences of equal-size numbers. Even this is difficult, because the values of the parameters of the problem that result in the difference of equal-size numbers often are not apparent, so that significant roundoff error cannot easily be predicted. The only practical resolution is to strive to write equations in forms that minimize the use of subtraction.

Relative Error

As already mentioned, the absolute error in the fixed-point number system is fixed, while the relative error in a floating-point number system is fixed. That is, the difference between two numbers in the fixed-point number system is always the same; in the floating-point number system it is not. The difference between two floating-point numbers, when the numbers are close to zero, is smaller than the difference between two floating-point numbers when the numbers are close to the maximum size in the calculator. The difference between two numbers divided by either of the numbers is approximately fixed in the floating-point number system, while it varies in the fixed -point number system. Thus the floating-point number system tends to emphasize relative rather than absolute error, as do most engineering and scientific analyses. Hence it is the natural number system for scientific calculations.

A similar situation occurs in evaluating functions. Scientific pocket calculator analysis favors functions written in a form that minimizes relative error rather than absolute error. Although this is well known to the experienced analyst, and seems quite rational to the practical analyst, we still find it prevalent to use "absolute error" as an accuracy criterion in numerical analysis. For example, calculating e^{-x} over the range 0 to 3, to 1 part in 10^3 using a Taylor series expansion, requires on the order of 12 terms in the series. That is, the contribution made by the thirteenth term in the Taylor series expansion of e^{-x}, when $x = 3$, is something less than 10^{-3}. However, if, more reasonably, we require that the relative error be 1 part in 10^3, only nine terms are needed in the series. The absolute error criterion requires 30% more terms than what is usually required for engineering analysis. In general, when deriving approximation formulas, it is important to decide what type of error is important to the problem being solved and use approximation methods that provide the appropriate accuracy. Too often the approximation is laboriously long and too accurate for the purpose.

In Chapter 4 we evaluate power series forms of advanced mathematical functions such as Bessel's functions and Legendre polynomials. Our emphasis then in these cases, is on both relative error and absolute error. The formulas based on relative error criteria have fewer terms than those based on absolute error. In scientific work (where relative error is of concern) this results in a significant reduction in the work required to evaluate these functions on the pocket calculator because the number of key strokes involved in raising the argument to high powers is eliminated.

Rearranging Expressions to Minimize Error in Function Evaluation

The pocket calculator's sole function is the numerical evaluation of mathematical functions. Hence it does not have alpha-numeric displays or the ability to display words, except by coincidence.* Its purpose is to evaluate functions. We have already examined the effects of underflow and overflow, roundoff, and the error criterion itself on the accuracy of numerical evaluation. Now we briefly look at the functions to be evaluated and how they may be written in forms where loss of accuracy due to the subtraction of two almost equal-sized numbers does not occur.

There are a number of "tricks" to handling the difference between two numbers that are close together. However, one general technique exists that can resolve many problem situations where the difference of two numbers are of approximately the same size. Consider the function

$$h(x) = f(x+\epsilon) - f(x)$$

As already discussed, the numerical evaluation of $h(x)$ can propagate roundoff error forward into the leading significant digits. This function can be modified as

$$h(x) = \{ f(x+\epsilon) - f(x) \} \left\{ \frac{f(x+\epsilon) + f(x)}{f(x+\epsilon) + f(x)} \right\}$$

$$h(x) = \frac{f^2(x+\epsilon) - f^2(x)}{f(x+\epsilon) + f(x)}$$

* See Appendix A.

This is a general equation that can, for algebraic and certain transcendental functions, transform the difference of two neighboring numbers into the ratio of sums of the numbers capable of being evaluated accurately on the pocket calculator (or on any calculator or computer). For example, if

$$f(x) = x^2$$

then

$$h(x) = \frac{(x+\epsilon)^4 - x^4}{2x^2 + 2x\epsilon + \epsilon^2} \cong \frac{4x^3\epsilon}{2x^2(1+\epsilon/x)} \cong 2x\epsilon$$

For another example, consider the function

$$f(x) = \sin(x+\epsilon)$$

Then

$$h(x) = \sin(x+\epsilon) - \sin(x) = 2\cos\left(x + \frac{\epsilon}{2}\right)\sin\left(\frac{\epsilon}{2}\right)$$

which for small ϵ (but not necessarily small x) is

$$h \cong 2\left\{\cos\left(x + \frac{\epsilon}{2}\right)\right\}\left\{\frac{\epsilon}{2}\right\} \cong \epsilon\cos\left(x + \frac{\epsilon}{2}\right)$$

An example suggested by Hamming is

$$(x+\epsilon)^{1/2} - (x)^{1/2} = \frac{\left[(x+\epsilon)^{1/2} - (x)^{1/2}\right]\left[(x+\epsilon)^{1/2} + (x)^{1/2}\right]}{(x+\epsilon)^{1/2} + (x)^{1/2}}$$

$$= \frac{\epsilon}{(x+\epsilon)^{1/2} + (x)^{1/2}}$$

With regard to other techniques, Hamming makes the interesting observation that what appear to be a large number of tricks to reformulate a function to handle its finite difference are really not new to the analyst. They are exactly the same methods used in calculus to derive the function's derivative. We can see this from the definition of the derivative

$$\lim_{\Delta x \to 0}\left\{\frac{\Delta y}{\Delta x}\right\} = \lim_{\Delta x \to 0}\left\{\frac{f(x+\Delta x) - f(x)}{\Delta x}\right\}$$

As a final resort to avoiding subtraction of nearly equal-sized numbers, most functions can be series expanded or approximated with different types of series for the interval of interest. Then $h(x)$ can be formed and modified as before to get around the subtraction problem.

An approach that works surprisingly well for certain functions (see Example 1-4) is to use the mean value theorem of differential calculus, where

$$f(b) - f(a) = (b - a)f'(\theta), \qquad (a < \theta < b)$$

As an example of the application of the mean value theorem, let us compute

$$h(x) = \sin(x + \epsilon) - \sin(x)$$

where $x + \epsilon$ is not necessarily small. Using the mean value theorem, we then find

$$h(x) = [(x + \epsilon) - (x)]\cos(\theta) = \epsilon \cos\theta$$

for

$$x + \epsilon < \theta < x$$

The difficulty is in selecting the value of θ that will accurately compute $f(x)$; that is, in selecting θ that produces less error than would be produced by the propagation of the roundoff error into the most significant digits. The author knows of no method for effectively estimating θ to ensure accuracy greater than is given by taking the difference itself. However, the midvalue interval is an obvious possibility. In this case, we find

$$h(x) \cong \epsilon \cos\left(x + \frac{\epsilon}{2}\right)$$

Clearly this method is of questionable value (for precision evaluation) except when θ can be determined. The equation is useful, however, for computing the extreme values of the difference by using the expressions

$$\epsilon \cos(x + \epsilon)$$

$$\epsilon \cos(x)$$

on the interval of the calculation.

A few commonly used difference equations for circumventing large errors in taking the difference between nearly equal values of popular transcendental functions are tabulated in Table 1-7.

Table 1-7 Commonly Used Difference Equations in Functional Evaluation

---

$$\Delta e^x = e^x (e^{\Delta x} - 1)$$

$$\Delta \ln(x) = \ln \left(1 + \frac{\Delta x}{x} \right)$$

$$\Delta \sin(2\pi x) = 2 \sin(\pi \Delta x) \cos \left[2\pi \left(x + \frac{\Delta x}{2} \right) \right]$$

$$\Delta \cos(2\pi x) = -2 \sin(\pi \Delta x) \sin \left[2\pi \left(x + \frac{\Delta x}{2} \right) \right]$$

$$\Delta \tan(2\pi x) = \sin(2\pi \Delta x) \sec(2\pi x) \sec(2\pi x + 2\pi \Delta x)$$

1-11 REFERENCES

One comprehensive, readable volume was selected to use as a reference throughout this book: Richard Hamming's *Numerical Methods for Scientists and Engineers*. This book, published by McGraw-Hill, is in its second printing (1973). Dr. Hamming has also published a superb textbook entitled *Introduction to Applied Numerical Analysis* (McGraw-Hill, 1971). For this chapter refer to Hamming's *Numerical Methods for Scientist and Engineers*, Chapters 2 and 3.

Example 1-1 Evaluate $\ln(0.9)$ using the fifth-order truncated Taylor series expansion of $\ln(1 + x)$ in the neighborhood of $x = 1$.

$$\ln(1 + x) \cong x \left(1 - \frac{x}{2} \left(1 - \frac{2x}{3} \left(1 - \frac{3x}{4} \left(1 - \frac{4x}{5} \right) \right) \right) \right), \qquad |x| < 1$$

Now

$$1 + x = 0.9$$

$$\therefore x = -0.1$$

Then

$$\ln(0.9) \cong -0.1 \left(1 - \frac{0.1}{2} \left(1 - \frac{2 \times 0.1}{3} \left(1 - \frac{3 \times 0.1}{4} \left(1 - \frac{4 \times 0.1}{5} \right) \right) \right) \right)$$

A typical algebraic key stroke sequence for evaluating this polynomial is

$$4 \times 0.1 \,\text{CHS} \div 5 + 1 \times 3 \times 0.1 \,\text{CHS} \div 4 + 1 \times 2 \times 0.1 \,\text{CHS} \div 3$$

$$+ 1 \times 0.1 \,\text{CHS} \div 2 + 1 \times 0.1 \,\text{CHS} =$$

A typical reverse polish key stroke sequence is

$$4\uparrow0.1\times5\div\text{CHS}\,1+3\times0.1\times4\div\text{CHS}1+2\times0.1\times3\div\text{CHS}\,1$$

$$+0.1\times2\div\text{CHS}\,1+0.1\times\text{CHS}$$

Accuracy considerations over a broader range of x are given in Table 1-8.

Table 1-8   Accuracy of the Fifth-Order Taylor Series Expansion of ln $(1 + x)$

$(1+x)$	x	$\ln(1+x)$	$x[1 - x/2(1 - \cdots)]$	Absolute Error	Relative Error (%)
0.9	−0.1	−0.10536052	−0.10536033	−0.00000018	00.000173
0.8	−0.2	−0.22314355	−0.22313067	−0.00001288	00.005774
0.7	−0.3	−0.35669494	−0.35651100	−0.00016394	00.04596194
0.6	−0.4	−0.51082562	−0.50978133	−0.00104429	00.20443190
0.5	−0.5	−0.69314718	−0.68854167	−0.00460551	00.80241261
0.4	−0.6	−0.91629073	−0.89995200	−0.01633873	01.78313839
0.3	−0.7	−1.20397280	−1.15297233	−0.05100047	04.23601512
0.2	−0.8	−1.60943791	−1.45860264	−0.15083525	09.37192077
0.1	−0.9	−2.30258509	−1.83012300	−0.47246209	20.51876799

Example 1-2   Evaluate $\ln(1 + x)$ using the fifth-order Chebyshev approximating polynomial

$$\ln(1+x) \cong x\big(a_1 + x(a_2 + x(a_3 + x(a_4 + a_5 x)))\big), \qquad 0 \leqslant x \leqslant 1$$

over the range $0 \leqslant x \leqslant 1$ using the coefficients (from page 55)

$$a_1 = 0.999949556 \qquad a_4 = -0.13606275$$
$$a_2 = -0.49190896 \qquad a_5 = 0.03215845$$
$$a_3 = 0.28947478$$

The accuracy of this approximation is shown in Tables 1-9 and 1-10.

Note that even outside the region where the approximating polynomial was designed to best approximate $\ln(1 + x)$ it is more accurate than the "unconditioned" Taylor series expansion of $\ln(1 + x)$. Evaluation of $\ln(1 + x)$ using the approximating polynomial requires approximately 60 key strokes (20 more keystrokes than are used in the Taylor series approximation) whether using the reverse-polish or algebraic languages. The addi-

Table 1-9   Accuracy of the Fifth-Order Chebyshev Polynomial Approximation of ln(1 + x)

$(1+x)$	x	$\ln(1+x)$	$x[a_1 + x(a_2 + \cdots)]$	Absolute Error	Relative Error (%)
1.1	+0.1	0.09531018	0.9530666	0.00000352	0.003697
1.2	+0.2	0.18232156	0.18233114	0.00000959	0.005257
1.3	+0.3	0.26236426	0.26236872	−0.00000445	−0.001697
1.4	+0.4	0.35647224	0.33646527	0.00000696	0.002070
1.5	+0.5	0.40546511	0.40545592	0.00000919	0.002267
2.0	+1.0	0.69314718	0.69315708	0.00000990	−0.001428

Table1.10   Accuracy of the Fifth-Order Chebyshev of ln(1 + x) Outside the Design Range of the Chebyshev Approximation

$1+x$	x	$\ln(1+x)$	$x[a_1 + x(a_2 + \cdots)]$	Absolute Error	Relative Error (%)
0.9	−0.1	−0.10536052	−0.10517205	−0.00018847	0.178879
0.8	−0.2	−0.22314355	−0.22211926	−0.00102429	0.459028
0.7	−0.3	−0.35667494	−0.35311655	−0.00355840	0.997658
0.6	−0.4	−0.51082562	−0.50084255	−0.00998309	1.954301
0.5	−0.5	−0.69314718	−0.66841824	−0.02472894	3.567632
2.1	1.1	0.74193734	0.74210824	−0.00017089	0.023034
2.2	1.2	0.78845736	0.78913899	−0.00068162	0.086450
2.3	1.3	0.83290912	0.83478743	−0.00187831	−0.225512
2.4	1.4	0.87546874	0.87972822	−0.00425948	−0.486537
2.5	1.5	0.91629073	0.92481112	−0.00852039	−0.929878

tional key strokes are associated mainly with entering the coefficients a_1, $a_2, \ldots, a_5$.

Example 1-3   Rewrite the difference

$$h(x) = \frac{1}{x+1} - \frac{1}{x}$$

in a form that will minimize roundoff error using a series expansion technique. The objective is to eliminate the differencing of two numbers of approximately equal size. Expanding the first term, we see that

$$\frac{1}{x+1} = \frac{1/x}{1+1/x} = \frac{1}{x}\left(1 - \frac{1}{x} + \frac{1}{x^2} - \frac{1}{x^3} + \cdots\right). \qquad |x| > 1$$

Then

$$h(x) = \frac{1}{x}\left(1 - \frac{1}{x} + \frac{1}{x^2} - \frac{1}{x^3} + \cdots\right) - \frac{1}{x}$$

$$h(x) = \frac{1}{x}\left[\left(1 - \frac{1}{x} + \frac{1}{x^2} - \frac{1}{x^3} + \cdots\right) - 1\right]$$

$$h(x) = -\frac{1}{x^2}\left(1 - \frac{1}{x} + \frac{1}{x^2} - \cdots\right)$$

$$h(x) = \frac{-1}{x^2}\left(\frac{1}{1 + 1/x}\right) = \frac{-1}{x(x+1)}$$

This form of $h(x)$ does not involve computing the difference of two numbers of nearly equal size. The range of x over which this derivation applies is $|x| > 1$.

Example 1-4   Rewrite the difference

$$h(x) = \frac{1}{x+1} - \frac{1}{x}$$

in a form that will minimize roundoff error using algebra.
   Cross-multiplying, we find

$$h(x) = \frac{x - (x+1)}{x(x+1)} = \frac{-1}{x(x+1)}$$

This result is the same as that developed with the series expansion method except that it holds for all x, not just $|x| > 1$. This is an important point to remember. Derivations using series expansion techniques *often* lead to results that hold over a greater range of the independent variable than their derivation strictly allows. With a pocket calculator it is easy to check the dynamic range over which a derived formula will work.

Example 1-5   Estimate $\sin(31°) - \sin(30°)$ using the mean value theorem. By the mean value theorem, we obtain

$$h(x) = \sin(30° + 1°) - \sin(30°) \approx 0.017453293 \cos(30.5°)$$

Here 0.017453293 is the value of 1° in radians. Then:

$$0.017453293 \cos(30.5°) = 0.015038266$$

$$\sin(31°) - \sin(30°) = 0.015038075$$

$$\text{relative error } (\%) = -0.0012700$$

$$\text{absolute error} = -0.000000191$$

Table 1-11 indicates that the mean value theorem can be useful for engineering evaluations, since the relative error is very small. Care must be taken, however, in using the mean value theorem. Had we used cos(30°) instead of cos(30.5°) we would find

$$0.017453293 \cos(30°) = 0.015114995$$

where actually

$$\sin(31°) - \sin(30°) = 0.015038075$$

$$\text{absolute error} = -0.000076920$$

$$\text{relative error } (\%) = 0.5115024$$

Table 1-11 Accuracy of Mean Value Theorem Approximation of $\sin(\theta + 1°) - \sin\theta$

θ (degrees)	Sin $(\theta + 1°) - \sin\theta$	Mean Value Theorem	Absolute Error	Relative Error (%)
0	0.017452406	0.017452628	−0.000060222	−0.0012693
10	0.017160818	0.017161036	−0.000000218	−0.0012693
20	0.016347806	0.016348014	−0.000000208	−0.0012604
30	0.015038075	0.015038266	−0.000000191	−0.0012700
40	0.013271419	0.013271588	−0.000000169	−0.0012708
50	0.011101518	0.011101659	−0.000000141	−0.0012723
60	0.008594304	0.008594412	−0.000000109	−0.0012629
70	0.005825955	0.005824029	−0.000000074	−0.0012695
80	0.002880587	0.002880624	−0.000000037	−0.0012756
90	0.000152305	0.000152307	−0.000000002	−0.0013714

Had we used $\cos(31°)$ instead of $\cos(30.5°)$, we would find

$$0.017453293 \cos(31°) = 0.014960392$$
$$\sin(31°) - \sin(30°) = 0.015038075$$
$$\text{absolute error} = 0.000077683$$
$$\text{relative error } (\%) = 0.5165751$$

Here we see that the relative error at the boundaries of the θ interval has jumped from $\sim \frac{1}{1000}\%$ when θ is taken at the midvalue of the interval to $\sim \frac{1}{5}\%$ when θ is taken at the end value of the interval.

CHAPTER 2

DIFFERENCE TABLES, DATA ANALYSIS, AND FUNCTION EVALUATION

2-1 INTRODUCTION

This chapter deals with interpolation, extrapolation, and smoothing of tabulated data. Many books on numerical analysis discuss these topics as related to the use of mathematical tables. Though we are interested in the use of these methods for precision table lookup, this chapter aims mainly to develop functions that are simple in form that can be used to replace complex functions. This technique, called analytic substitution, is commonplace in advanced analysis. For example, cost data developed on computer programs with as many as 500 cost-estimating relationships (CERs) can be used to generate a table of costs as a single design parameter is changed. It is often convenient to develop an interpolation formula based on the table of discrete costs which will compute system cost as a function of the single design parameter. The simpler formula can be analytically substituted for the entire complex system of CERs in the large-scale cost model. This reduces the cost of "cost estimating" and makes the simplified models convenient to analyze on the pocket calculator (see Chapter 11). We will also investigate the smoothing of tabulated data on the basis of estimates of the error propagated in a difference table. Finally, we will study what is perhaps the most important but seemingly least developed use of data tables, extrapolation or prediction. Here projections, predictions, and identification of trends and predicted values of function are discussed both from the viewpoints of mathematical limitations and the practical necessity to predict the behavior of dynamic processes from their data tables.

2-2 DIFFERENCE TABLES OF EQUALLY SPACED DATA

Before the age of pocket calculators, the preparation of extensive difference tables of data with mechanical calculators was laborious and noisy at best, and frustrating at worst. In practical analysis they require carrying numbers to at least as many as five significant digits. A table of finite differences of n numbers and m differences requires

$$\frac{m(2n-m-1)}{2}$$

differences to be calculated and recorded in the difference table. For a table of 50 entries and 5 differences, this involves 235 differences to be computed. Thus 470 data entries must be made, which took about an hour on the old mechanical calculators. On the electronic pocket calculator these calculations are done quickly and quietly, with the time-limiting element being the analyst's preparation of the difference table. Tables of fifth-order differences of 50 numbers can be conveniently prepared in approximately 15 minutes with any pocket calculator.

The difference tables that we are concerned with here are usually generated in two ways. Either a function is evaluated for certain values of its independent variable or data are determined by measurement of an experiment. In both cases the tables of equally spaced data can usually be prepared, especially of data determined from experiments, since much of experimental electronics and data sampling is done digitally and can be time-referenced to a digital clock. We discuss arbitrarily spaced data later in this chapter.

Our notation is based exclusively on the definition of the forward difference:

$$\Delta y_i = y_{i+1} - y_i = y(x_0 + [i+1]\Delta x) - y(x_0 + i\Delta x), \qquad i = 0, 1, 2, \ldots, n$$

Figure 2-1 illustrates the definitions of the differences involved in the difference table. Occasionally we use the term h to represent the spacing of the data, that is, $\Delta x = h = x_{n+1} - x_n$. We do not use backward differences or central differences in this book. Backward and central differences are only useful for changing the form of equations used in the derivation of numerical approximation methods. Since our interest here is not in manipulating equations but in their numerical evaluation, we use only the forward difference notation. Repeated application of the definition of the forward difference generates higher-order differences. For example, the

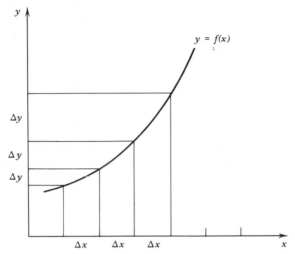

Figure 2-1   Definition of differences of equally spaced data.

second-order difference is derived as

$$\Delta^2 y_i = \Delta y_{i+1} - \Delta y_i$$

$$= y_{i+2} - y_{i+1} - (y_{i+1} - y_i)$$

$$= y_{i+2} - 2y_{i+1} + y_i$$

The third-order difference is developed as

$$\Delta^3 y_i = \Delta^2 y_{i+1} - \Delta^2 y_i$$

$$= \Delta y_{i+2} - \Delta y_{i+1} - (\Delta y_{i+1} - \Delta y_i)$$

$$= (y_{i+3} - y_{i+2}) - (y_{i+2} - y_{i+1}) - (y_{i+2} - y_{i+1}) + (y_{i+1} - y_i)$$

$$= y_{i+3} - 3y_{i+2} + 3y_{i+1} - y_i$$

The differences can be numerically evaluated using the equations just developed, or they can be computed directly from the tabulated values of the dependent variable, as shown in Figure 2-2.

The nth difference operator is given by the formula

$$\Delta^n = (z-1)^n = z^n - nz^{n-1} + \frac{n(n-1)}{2} z^{n-2} - \cdots \qquad (2\text{-}1)$$

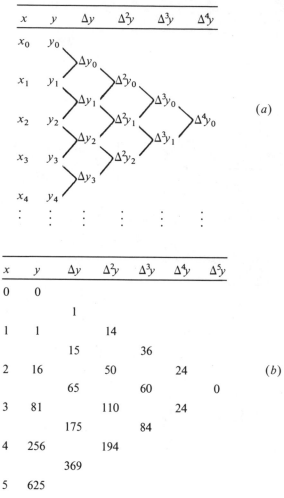

Figure 2-2 Finite difference tables. (*a*) Difference table definition. (*b*) Numerical example $y = x^4$.

where z is the shifting operator defined by the relation

$$z[y(z)] = y(x + \Delta x)$$

Furthermore, by repeated application of the shifting operator we see that

$$z^n[y(x)] = y(x + n\Delta x)$$

Equation 2-1 is derived by noting that the forward difference and shifting operators are related as follows:

$$\Delta y_i \quad = \quad y_{i+1} - y_i = zy_i - y_i = (z-1)y_i$$

$$\vdots \qquad \vdots \qquad\qquad\qquad \vdots$$

$$\Delta^n y_i \quad = \qquad\qquad\qquad (z-1)^n y_i$$

Note also that equation 2-1 can be written in the form

$$\Delta^n y_i = [z^n - C(n,1)z^{n-1} + C(n,2)z^{n-2} - \cdots]y_i$$

$$\Delta^n y_i = y_{i+n} - C(n,1)y_{i+n-1} + C(n,2)y_{i+n-2} - \cdots$$

where

$$C(n,m) = \frac{n!}{m!(n-m)!}$$

which is the mth binomial coefficient of order n.

2-3 DATA INTERPOLATION

Armed with these definitions, we are now prepared to examine a number of formulas for analytic substitution or for interpolation. The method that we use here involves a Lozenge diagram of differences and binomial coefficients which can be combined into interpolation formulas. The diagram is shown in Figure 2-3. Certain rules applied along paths across the diagram proceeding from left to right define interpolation formulas. This diagram is so general that it encompasses both Newton's forward and backward difference formulas, Stirling's interpolation formula, Bessel's interpolation formula, and an interesting and unusual formula due to Gauss which zigzags across the diagram. The rules to be followed that generate these and many more interpolation formulas are the following:

1. When moving from left to right across the diagram, sum at each step.
2. When moving from right to left across the diagram, subtract at each step.
3. If the slope of the step is positive, the term in the interpolation formula for that step is the product of the difference crossed times the factor immediately below it.
4. If the slope of the step is negative, the term is the product of the difference crossed times the factor immediately above it.
5. If the step is horizontal and passes through a difference, the term is the product of the difference times the average of the factors above and below it.

6. If the step is horizontal and passes through a factor, the term is the product of the factor times the average of the differences above and below it.

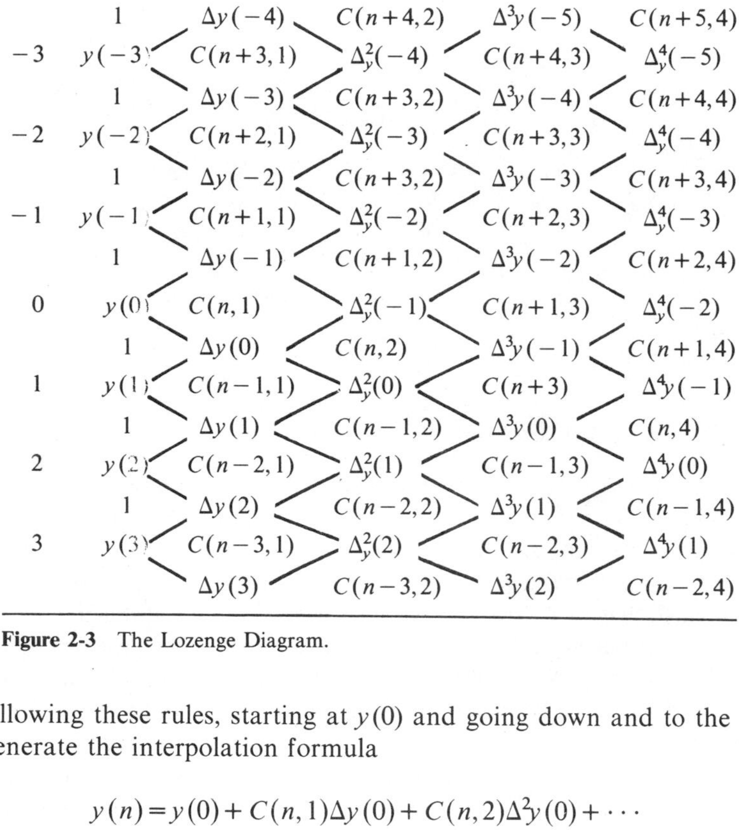

	1	$\Delta y(-4)$	$C(n+4,2)$	$\Delta^3 y(-5)$	$C(n+5,4)$
-3	$y(-3)$	$C(n+3,1)$	$\Delta_y^2(-4)$	$C(n+4,3)$	$\Delta_y^4(-5)$
	1	$\Delta y(-3)$	$C(n+3,2)$	$\Delta^3 y(-4)$	$C(n+4,4)$
-2	$y(-2)$	$C(n+2,1)$	$\Delta_y^2(-3)$	$C(n+3,3)$	$\Delta_y^4(-4)$
	1	$\Delta y(-2)$	$C(n+3,2)$	$\Delta^3 y(-3)$	$C(n+3,4)$
-1	$y(-1)$	$C(n+1,1)$	$\Delta_y^2(-2)$	$C(n+2,3)$	$\Delta_y^4(-3)$
	1	$\Delta y(-1)$	$C(n+1,2)$	$\Delta^3 y(-2)$	$C(n+2,4)$
0	$y(0)$	$C(n,1)$	$\Delta_y^2(-1)$	$C(n+1,3)$	$\Delta_y^4(-2)$
	1	$\Delta y(0)$	$C(n,2)$	$\Delta^3 y(-1)$	$C(n+1,4)$
1	$y(1)$	$C(n-1,1)$	$\Delta_y^2(0)$	$C(n+3)$	$\Delta^4 y(-1)$
	1	$\Delta y(1)$	$C(n-1,2)$	$\Delta^3 y(0)$	$C(n,4)$
2	$y(2)$	$C(n-2,1)$	$\Delta_y^2(1)$	$C(n-1,3)$	$\Delta^4 y(0)$
	1	$\Delta y(2)$	$C(n-2,2)$	$\Delta^3 y(1)$	$C(n-1,4)$
3	$y(3)$	$C(n-3,1)$	$\Delta_y^2(2)$	$C(n-2,3)$	$\Delta^4 y(1)$
		$\Delta y(3)$	$C(n-3,2)$	$\Delta^3 y(2)$	$C(n-2,4)$

Figure 2-3   The Lozenge Diagram.

Following these rules, starting at $y(0)$ and going down and to the right, we generate the interpolation formula

$$y(n) = y(0) + C(n,1)\Delta y(0) + C(n,2)\Delta^2 y(0) + \cdots$$

which becomes

$$y(n) = y(0) + n\Delta y(0) + \frac{n(n-1)}{2}\Delta^2 y(0) + \cdots$$

This is Newton's forward difference interpolation formula. To generate Newton's backward difference formula, the procedure is reversed. Starting at $y(0)$ and moving up and to the right, we generate the formula

$$y(n) = y(0) + C(n,1)\Delta y(-1) + C(n+1,2)\Delta^2 y(-2) + \cdots$$

which becomes

$$y(n) = y(0) + n\Delta y(-1) + \frac{n(n+1)}{2}\Delta^2 y(-2) + \cdots$$

This is Newton's backward difference formula.

To develop Stirling's formula, we start at $y(0)$ and move horizontally to the right. In this case, we generate the interpolation formula

$$y(n) = y(0) + C(n,1)\left\{\frac{\Delta y(0) + \Delta y(-1)}{2}\right\}$$

$$+ \left\{\frac{C(n+1,2) + C(n,2)}{2}\right\}\Delta^2 y(-1) + \cdots$$

$$y(n) = y(0) + n\left\{\frac{\Delta y(0) + \Delta y(-1)}{2}\right\} + \frac{n^2}{2}\Delta^2 y(-1) + \cdots$$

Bessel's formula can be generated by starting midway between $y(0)$ and $y(1)$.

$$y(n) = 1\left\{\frac{y(0) + y(1)}{2}\right\} + \left\{\frac{C(n,1) + C(n-1,1)}{2}\right\}\Delta y(0) + \cdots$$

$$y(n) = \left\{\frac{y(0) + y(1)}{2}\right\} + (n - \tfrac{1}{2})\Delta y(0)$$

$$+ \frac{n(n-1)}{2}\left\{\frac{\Delta^2 y(-1) + \Delta^2 y(0)}{2}\right\} + \cdots$$

Clearly, a great number of other formulas can be generated and used for interpolation of data.

Interpolation is often employed in computing intermediate values of tabulated functions. While the scientific pocket calculator gives sine, cosine, tangent, arc sine, arc cosine, and arc tangent (and, for some of the more advanced scientific machines, hyperbolic sine, hyperbolic cosine, and hyperbolic tangent), they usually do not have the capability of generating Bessel's functions, Legendre polynomials, error functions, and the like. Those are often more easily evaluated with standard reference tables. In these cases, it is occasionally necessary to interpolate between two values in the table.

Before discussing the interpolation process, however, it is worth pointing out that most well-made tables are often generated with auxiliary functions as opposed to the actual functions themselves. For example, the exponential integral with positive argument is given by

$$Ei(x) = \int_{-\infty}^{x} \frac{e^u}{u} \, du$$

which takes the series form

$$Ei(x) = \gamma + \ln(x) + \frac{x}{1 \cdot 1!} + \frac{x^2}{2 \cdot 2!} + \frac{x^3}{3 \cdot 3!} + \cdots$$

can be approximated with the series

$$Ei(x) \cong \frac{e^x}{x} \left[1 + \frac{1!}{x} + \frac{2!}{x^2} + \frac{3!}{x^3} + \cdots \right] \qquad (x \to \infty)$$

The logarithmic singularity in the first series does not permit easy interpolation near $x = 0$. The function $Ei(x) - \ln x$ is better behaved and more readily interpolated when x is near zero. In fact, $x^{-1}[Ei(x) - \ln(x) - \gamma]$ (where γ is Euler's constant $0.577 \cdots$) is an auxiliary function that results in a slightly higher interpolation accuracy than when $Ei(x)$ is computed from interpolated values of the table of $Ei(x)$ directly.

Generally tables are constructed and presented so that reasonable-order interpolating polynomials (i.e., first-, second-, or third-order) can be used to compute intermediate values while retaining the precision of the table.

For example, in the *Handbook of Mathematical Functions* (U.S. Department of Commerce, Bureau of Standards, Applied Mathematics Series 55) most tables are accompanied by a statement of the maximum error in a linear interpolation between any two numbers in the table, and the number of function values needed in Laplace's formula or Atkins' method to interpolate to nearly full tabular accuracy.

An example from the *Handbook of Mathematical Functions* appears in Table 2-1. The accuracy statement is given in brackets. The numbers in brackets mean that the maximum error in a linear interpolate is 3×10^{-6} and that to interpolate to the full tabular accuracy, five points must be used in Lagrange's method or Atkins' method of interpolation. The linear interpolation formula is

$$f_p = (1 - p)f_0 + pf_1$$

Table 2-1   Exponential Integral Auxiliary Function

x	$xe^xE_1(x)$	x	$xe^xE_1(x)$
7.5	0.892687854	8.0	0.898237113
7.6	0.893846312	8.1	0.899277888
7.7	0.894979666	8.2	0.900297306
7.8	0.896088737	8.3	0.901296023
7.9	0.897174302	8.4	0.902274695

$$\begin{bmatrix} (-6)3 \\ 5 \end{bmatrix}$$

where f_0, f_1 are consecutive tabular values of the function corresponding to arguments x_0, x_1 respectively; p is the given fraction of the argument interval

$$p = \frac{(x - x_0)}{(x_1 - x_0)}$$

and f_p is the required interpolate. For example, if we interpolate between the values of Table 2-1 for $x = 7.9527$, we find that

$$f_0 = 0.897174302$$

$$f_1 = 0.898237113$$

$$p = 0.527$$

We then obtain

$$f_{0.527} = (1 - 0.527)(0.897174302) + 0.527(0.898237113)$$

$$f_{0.527} = 0.897734403.$$

The terms in the brackets indicated that the accuracy for linear interpolation was 3×10^{-6}. Thus we round this result to 0.89773. The maximum possible error in this answer is composed of the error committed by the last rounding, that is, $0.4403 \times 10^{-5} + 3 \times 10^{-6}$, and thus certainly cannot exceed 0.8×10^{-5}.

To get greater precision, we can interpolate this example of the table using Lagrange's formula. In this example, the interpolation formula is the five-point formula:

$$f(x_0 + p\Delta x) = \left\{\frac{(p^2-1)(p-2)p}{24}\right\}f_{-2} - \left\{\frac{(p-1)(p^2-4)p}{6}\right\}f_{-1}$$

$$+ \left\{\frac{(p^2-1)(p-2)p}{4}\right\}f_0 - \left\{\frac{(p+1)(p^2-4)p}{6}\right\}f_1$$

$$+ \left\{\frac{(p^2-1)(p+2)p}{24}\right\}f_2, \qquad |p| < 1$$

Another approach is to use a five-term Newton forward or backward difference formula, a Bessel's formula, Stirling's formula, or any of the formulas that come out of the Lozenge diagram. The details associated with such interpolations are conveniently found in Chapter 25 of the *Handbook of Mathematical Functions.*

Since there are occasions for using inverse interpolation, we discuss it briefly here. If we are given a table of values of the dependent variable y_n as a function of values of the independent variable x_n,

$$y_n = f(x_n) \qquad \text{(tabulated function)}$$

then intermediate values of y can be computed by interpolating between the values y_n with an interpolating polynomial $g(x)$ as

$$y = g(x) \cong f(x) \qquad \text{(continuous function)}$$

Inverse interpolation is a matter of viewpoint. Here we would view the interpolation from the standpoint of the dependent variable,

$$x_n = f^{-1}(y_n) \qquad \text{(tabulated function)}$$

Then intermediate values of x can be computed by interpolating between the values of x_n with an interpolating polynomial $h(y)$ as

$$x = h(y) \cong f^{-1}(y) \qquad \text{(continuous function)}$$

With linear interpolation there is no difference in principle between direct and inverse interpolation. In cases where the linear formula is not sufficiently accurate, two methods are available for accuracy improvement. The first is to interpolate more accurately by using, for example, a higher-order Lagrange's formula or an equivalent higher-order polynomial method. The second is to prepare a new table with a smaller interval in the neighborhood of interest, and then apply accurate inverse linear interpolation to the subtabulated values.

It is important to realize that the accuracy of inverse interpolation may be very different from that of a direct interpolation. This is particularly true in regions where the function is slowly varying, such as near flat maximum or minimum. The maximum absolute error resulting from inverse interpolation can be estimated with the aid of the formula

$$\delta x = \left(\frac{\partial f}{\partial x} \right)^{-1} \delta y, \quad \delta x \approx \left(\frac{\Delta f}{\Delta x} \right)^{-1} \delta y$$

where δy is the maximum possible error in the tabulation of y values and Δf and Δx are the first differences generated from the table in the neighborhood of the region of interest.

Let us now return our attention to the generation of difference tables. In the generation of interpolating polynomials of reasonable size, the finite differences in the difference table must be small for high-order differences. If they are not small the questions is, "What can we do to reduce the size of the finite differences?"

There are only three considerations associated with any difference table. The first is the number of differences to which the table is taken the second is the spacing between different values of the tabulated function, and the third is the number of figures tabulated. The effect of halving (factor of $\frac{1}{2}$) spacing in the independent variable x is to divide the first differences by 2, the second differences by 4, the third differences by 8, and so on. Examples of the effect that different spacings of x have on the function $y = x^3$ appear in Table 2-2. In answer to the question above then: to reduce the nth-order difference by a factor k we must reduce the data interval (independent variable) by a factor of approximately $\sqrt[n]{k}$.

2-4 DATA EXTRAPOLATION

Extrapolation outside the range of data that makes up a difference table is a controversial procedure. It is, however, a procedure of great practical interest. Given the behavior of a dynamic process, sampled at intervals, it is only natural to ask to what extent the table can be extended beyond the range of the data used to make up the table to *predict* the future behavior of the process being considered. This is a very practical, important, and real matter. It is the problem of science to be interested in predicting the behavior of systems based on observations of their past behavior. While there are a number of stock market "chartsmen" who use finite difference techniques, it is generally accepted that extrapolation outside the range of the difference table is as much an art as a science. Because of its practical

Table 2-2 The Effect of Interval Halving on The Finite Differences in a Difference Table[a]

Full interval $\Delta x = 2$					Half-interval $\Delta x = 1$				
x	$y = x^3$	Δy	$\Delta^2 y$	$\Delta^3 y$	x	$y = x^3$	Δy	$\Delta^2 y$	$\Delta^3 y$
0	0				0	0			
		8					1		
2	8		48		1	1		6	
		56		48			7		6
4	64		96		2	8		12	
		152		48			19		6
6	216		144		3	27		18	
		296		48			37		6
8	512		192		4	64		24	
		488					61		
10	1000				5	125			

[a]Third-order difference is reduced by a factor of 8 when interval between values of x is halved.

value and practical interest, it will be covered here but with the proviso that the reader recognize that extrapolation is a questionable procedure. That is, the same difference table using only slightly different extrapolation techniques can, and usually does, lead to significantly different predictions. Because of this lack of robustness of extrapolated data, the procedure has questionable value.

We illustrate the problem of prediction with the following practical example. Consider an aircraft executing a fully automatic landing. Sampled values of the altitude are shown in Table 2-3. What will be the conditions at touchdown? This particular example is a nontrivial one, in that the heart of present-day flight-control performance monitors hinges on the ability to predict the dynamic behavior of high-energy devices, such as aircraft, when terminal operations are under automatic control. The obvious first step is to form the difference table, as shown in Table 2-3. We note that this table can be carried to the third difference without the lower-order differences becoming constant. The obvious next step to extrapolation is to assume that the third difference holds constant up to touchdown, and to predict the behavior shown in Table 2-4. Clearly the predicted results are fairly grim. We have low confidence in extrapolations of this type because the difference table did not indicate the influence of any control law through arriving at constant differences between any of the finite differences. Had we found, for example, that all of the second differences held constant and the third differences were zero, we might be

Table 2-3   Difference Table of Altitude of an Aircraft Executing an Automatic Landing

t	$h(t)$	Δh	$\Delta^2 h$	$\Delta^3 h$
0	60			
		-13		
1	47		$+3$	
		-10		-2
2	37		$+1$	
		-9		0
3	28		$+1$	
		-8		
4	20			

entitled to higher confidence in the extrapolation to touchdown by assuming that the guidance law objective was to hold the third-order differences to zero. It follows, then, that a procedure for increasing the confidence in extrapolation from finite difference tables is to find a transformation of the variable of interest which would expose the guidance law and its effect on the difference table.

Table 2-4   Extrapolated Touchdown Conditions of Automatically Landed Aircraft

	t	$h(t)$	Δh	$\Delta^2 h$	$\Delta^3 h$
	0	60			
			-13		
	1	47		$+3$	
			-10		-2
Range of actual data	2	37		$+1$	0 } Average $= -1$
			-9		
	3	28		$+1$	
			-8		-1
	4	20		0	
			-8		-1
	5	12		-1	
			-9		-1
Range of extrapolation	6	Touchdown sink rate	3	-2	
			-11		
	7	~ 11 fps[a] (hard)	-8		

[a] fps = feet per second.

After a little thought, it might be expected that the guidance law to automatically land an aircraft would be an exponential law of the form

$$\frac{dh}{dt} = -k(h + h_B)$$

which results in a flared landing path of the form

$$h = ho^{e^{-kt}} - h_B$$

which is sketched in Figure 2-4. This suggests the formation of the difference Table 2-5. Note that the second difference is near zero so that for short-term prediction (next 7 seconds) a reasonable assumption is that $\Delta \ln(h)$ is approximately constant and equal to -0.3062. The differences between the two approaches are tabulated in Table 2-6. It is apparent that the logarithmic extrapolation does better short-term prediction than does the "third-order" extrapolation.

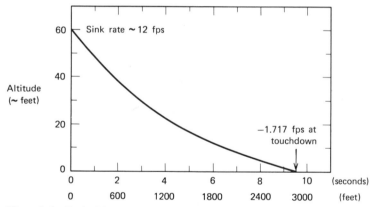

Figure 2-4   Typical jet transport landing trajectory (fps = feet per second).

In summary, we might expect to extrapolate with greater confidence from Table 2-5 than Table 2-4 because of the observed characteristics of the guidance law. From a strictly mathematical viewpoint, the issue is not all that clear. The number of actual samples of the second difference is small and thus the true mathematical confidence in the fact that the guidance law is in some way holding the second difference to zero is low. In the end, extrapolation using difference tables involves careful judgment.

Table 2-5 Logarithmic Extrapolation of Touchdown Conditions of Automatically Landed Aircraft

	t	$h(t)$	$\ln(h(t))$	Δh
Actual data range	0	60	4.0943	
				−0.4054
	1	47	3.8501	
				−0.2442
	2	37	3.6109	
				−0.2392
	3	28	3.3322	
				−0.3365
	4	20	2.9957	
				−0.3062
Range of extrapolation	5	14.7	2.6895	
				−0.3062
	6	10.84	2.3833	
				−0.3062
	7	7.98	2.0771	
				−0.3062
	8	5.876	1.7709	
				−0.3062
	9 ⎱ Sink rate	4.326	1.4647	
				−0.3062
	10 ⎰ ~1.141 fps[a]	3.185	1.1585	
				−0.3062
	11	2.345	0.8523	

Average $\cong 0.3062$ (average of the actual data range differences)

[a] fps = feet per second.

Table 2-6 Comparison of Extrapolation Methods for Predicting Touchdown Conditions

	Actual	"Third-Order" Extrapolation		First-Order Logarithmic Extrapolation	
t	h	h	Absolute Error	h	Absolute Error
5	16	12	−4	15	−1
6	11	3	−8	11	0
7	7	−8	−15	8	+1
8	4			6	+2
9	1.6			4	+2.6
10	0			3	+3
11	0			2	+2

[a] fps = feet per second.

2-5 DATA ERROR LOCATION AND CORRECTION

Errors due to observations, calculation, measurement, or recording often occur in a table of numbers. These errors introduced into the calculation process are significantly magnified in the generation of ascending differences in the difference table. This can be seen in Table 2-7. It is apparent that the errors propagate and are distributed binomially (in any given difference the errors are weighted by binomial coefficients). It is also apparent that the error grows rapidly as it propagates into ascending orders of difference. For example, the error in Table 2-8 might be antici-pated by noting the form of the third difference. We see the pattern of signs $(+)$, $(-)$, $(+)$, $(-)$ indicative of error propagation. Also, note that the pattern of fourth difference is centered on $y = 17$. Furthermore, note

Table 2-7   Error Propagation in Difference Tables

y	Δy	$\Delta^2 y$	$\Delta^3 y$	$\Delta^4 y$
0				
	0			
0		0		
	0		0	
0		0		$+\epsilon$
	0		$+\epsilon$	
0		$+\epsilon$		-4ϵ
	$+\epsilon$		-3ϵ	
$+\epsilon$		-2ϵ		$+6\epsilon$
	$-\epsilon$		$+3\epsilon$	
0		$+\epsilon$		-4ϵ
	0		$-\epsilon$	
0		0		$+\epsilon$
	0		0	
0		0		
	0			
0				

Table 2-8 Unit Error Propagation in the Difference Table for the Function $y = x^2$

x	y	Δy	$\Delta^2 y$	$\Delta^3 y$	$\Delta^4 y$
0	0				
		1			
1	1		2		
		3		0	
2	4		2		−1
		5		1	
3	9		3		−4
		8		−3	
4	17		0		+6
		8		+3	
5	25		3		−4
		11		−1	
6	36		2		+1
		13		0	
7	49		2		
		15			
8	64				

that 6ϵ in Table 2-7 corresponds to 6 in Table 2-8; that is,

$$6\epsilon = 6$$

$$\epsilon = 1$$

Moreover, if the error in the values of y were of the form

$$y = x^2 + 5$$

we can expect in the fourth difference column to show an error of $6k$. Thus one-sixth of the fourth difference which is centered on the number in error is a measure of the error—which can then be subtracted from the column of y values. We might modify Table 2-8 by replacing 17 with $(17-1) = 16$, thus obtaining the difference table shown in Table 2-9. In general, then, data smoothing is done by:

1. Keeping an eye open for the $(+)$, $(-)$, $(+)$, $(-)$,$\cdots$ pattern in high-order differences that indicates error propagation.
2. Identifying the tabulated value on which the pattern is centered.

Table 2-9 Smoothed Data Table for the Function
$y = x^2$

x	y	Δy	$\Delta^2 y$	$\Delta^3 y$	$\Delta^4 y$
0	0				
		1			
1	1		2		
		3		0	
2	4		2		0
		5		0	
3	9		2		0
		7		0	
4	16		2		
		9			
5	25				

3. Equating observed error with its binomial error counterpart.
4. Solving for the error and appropriately modifying the data table.
5. Testing the table for elimination of the $(+), (-), (+), (-), \ldots$ pattern.

2-6 MISSING ENTRIES

Occasionally a difference table has a few missing entries in the dependent variable. Missing entries in the difference table can be estimated in several ways.

The simplest method is to examine the table and decide whether the points could be reasonably fit with a polynomial. For example, a data table with four points, one of which is unknown, might be fit with a second-degree polynomial. It is characteristic of difference tables that nth-order differences of polynomials of degree $n-1$ equal zero. For example, the equation

$$y = 2x^2 + x + 3$$

has the difference equation shown in Table 2-10, where it is apparent that the third-order differences equal zero. This characteristic is present in general in nth-order polynomials; that is, their $(n+1)$st-order (and all higher) differences equal zero. Using this property, we would expect that the fourth-order difference would equal zero; that is

$$\Delta^4(y) = 0$$

Table 2-10   Difference Table for $y = 2x^2 + x + 3$

Subscript in Missing Entry Formula	x	y	Δy	$\Delta^2 y$	$\Delta^3 y$
	0	3			
			3		
	1	6		4	
			7		0
0	2	13		4	
			11		0
1	3	24		4	
			15		0
2	4	39		4	
			19		0
3	5	58		4	
			23		0
4	6	81		4	
			27		
	7	108			

This can be rewritten in the shifting-operator notation as

$$(z - 1)^4 y = (z^4 - 4z^3 + 6z^2 - 4z + 1)y = 0$$

This gives us

$$y_4 - 4y_3 + 6y_2 - 4y_1 + y_0 = 0$$

$$y_2 = \tfrac{1}{6}[4(y_1 + y_3) - (y_4 + y_0)]$$

Here we use an even-order difference because all even-order difference equations do the following:

1. Give one middle term, which can be centered in the missing number in the table.

2. Result in missing entry determination with a minimum of roundoff error.

3. Are numerically more stable than their odd-order counterparts.

Let us assume, for the sake of the discussion, that the $y = 39$ entry is missing in Table 2-10. We can substitute directly from the table with the missing data point to obtain

$$y_2 = \tfrac{1}{6}[4(24 + 58) - (81 + 13)]$$

from which we can solve for the missing data point:

$$y_2 = \tfrac{1}{6}[4(82) - 94] = \tfrac{1}{6}(328 - 94 = 234) = \tfrac{234}{6} = 39$$

The method just described for filling in missing values in the data table is particularly suited to analysis on the pocket calculator in that it does not involve the determination of unknown coefficients in a polynomial (the usual methods for missing data determination). For tables with large numbers, the arithmetic could be tedious, but with the pocket calculator it is a simple matter to perform the sums and products for tables of large values requiring high precision. Another point worth making regarding identification of missing entries in data tables is that for tables with large numbers of values, say on the order of 20 to 100, it is not necessary to look for twentieth-order differences to develop the formula for computing the missing data. One need only determine the polynomial that can be reasonably expected to fit locally through two, four, or six data points symmetrically placed about the missing value to find the missing point.

We have been stressing the determination of interpolating polynomials by way of finite difference tables because the pocket calculator enables one to find finite differences quickly and conveniently, thus leading immediately to interpolation formulas of high order and high accuracy, which themselves can be evaluated on the pocket calculator conveniently and to high precision. This, in fact, is the reason for using the pocket calculator with difference tables: high-order difference tables lead to high-order approximating polynomials, which, when written in nested parenthetical form, are easily evaluated on the pocket calculator to high precision.

The difficulty in using low-order polynomials for manual analysis in the precalculator era was that they generally were not sufficiently accurate to permit the precision numerical evaluation necessary for most engineering, economic, chemical,and other types of precision analysis. On the pocket calculator we can conduct precision analysis relatively quickly and efficiently by using high-order polynomials generated simply with difference tables of high order.

2-7 LAGRANGE'S INTERPOLATION FORMULA

So far we have studied the interpolation of equally spaced data through the use of difference tables and the Lozenge diagram as a convenient means for remembering a large number of different interpolation formulas. These interpolation formulas, however, do not apply to nonequally spaced values of the independent variable nor when the nth differences of the dependent variable are not small or zero. In these cases we can then use Lagrange's interpolation formula to develop a polynomial that can be used for analytic substitution. Though there are other interpolation formulas for unequally spaced data, the advantage to using Lagrange's interpolation formula is that the coefficients are particularly easy to remember, and to determine, with the pocket calculator. The method works for both nonequally and equally spaced data and regardless of whether the nth differences are small. Lagrange's interpolation formula is

$$y = y_0 \frac{(x-x_1)(x-x_2)\cdots(x-x_p)}{(x_0-x_1)(x_0-x_2)\cdots(x_0-x_p)} + y_1 \frac{(x-x_0)(x-x_2)\cdots(x-x_p)}{(x_1-x_0)(x_1-x_2)\cdots(x_1-x_p)}$$

$$+ \cdots + y_p \frac{(x-x_0)(x-x_1)\cdots(x-x_{p-1})}{(x_p-x_0)(x_p-x_1)\cdots(x_p-x_{p-1})}$$

An interesting and important feature of Lagrange's interpolation formula is that, if the data table has n entries, the formula appears to have n terms. It turns out, however, that if the table amounts to four or five samples of, say, a second-order polynomial, the terms will cancel, giving only the pieces due to the quadratic function. As an example of this, consider the data table shown in Table 2-11. Using Lagrange's interpolation formula, we have

$$y = 6 \frac{(x-5)(x-7)(x-9)(x-11)}{(3-5)(3-7)(3-9)(3-11)} + 24 \frac{(x-3)(x-7)(x-9)(x-11)}{(5-3)(5-7)(5-9)(5-11)}$$

$$+ 58 \frac{(x-3)(x-5)(x-9)(x-11)}{(7-3)(7-5)(7-9)(7-11)} + 108 \frac{(x-3)(x-5)(x-7)(x-11)}{(9-3)(9-5)(9-7)(9-11)}$$

$$+ 174 \frac{(x-3)(x-5)(x-7)(x9)}{(11-3)(11-5)(11-7)(11-9)}$$

The reader can now simplify the equation. It will be found that

$$y = 2x^2 - 7x + 9$$

which is a polynomial of degree *two*, rather than of the fourth degree, as might be expected from the fourth-order polynomials in the numerators of all the terms in Lagrange's interpolation formula. Because of roundoff the exact cancelation of the coefficients for the higher powers of x will not occur, but they will be very small, indicating that they should be made zero.

Table 2-11   Five Evaluations of a Quadratic
Equation

x	3	5	7	9	11
y	6	24	58	108	174

2-8 DIVIDED DIFFERENCE TABLES

Another approach to the generation of interpolation formulas for tables of data of unequally spaced values of the independent variable is to prepare a table of divided differences. Assuming that the values of the independent variable x are $x_0, x_1, x_2, x_3 \cdots$ and that the value of the dependent variable is $y = f(x)$, we can prepare a table of successive divided differences of the form

$$f(x_0, x_1) = \frac{f(x_1) - f(x_0)}{x_1 - x_0}$$

$$f(x_0, x_1, x_2) = \frac{f(x_1, x_2) - f(x_0, x_1)}{x_2 - x_0}$$

$$f(x_0, x_1, x_2, x_3) = \frac{f(x_1, x_2, x_3) - f(x_0, x_1, x_2)}{x_3 - x_0}$$

These terms are commonly called divided differences of orders $1, 2, 3$, and so on. We can now prepare a table of divided differences, as shown in Table 2-12,

Table 2-12   Divided Difference Table

$x(0)$				
	$f(x_0, x_1)$			
$x(1)$		$f(x_0, x_1, x_2)$		
	$f(x_1, x_2)$		$f(x_0, x_1, x_2, x_3)$	
$x(2)$		$f(x_1, x_2, x_3)$		$f(x_0, x_1, x_2, x_3, x_4)$
	$f(x_2, x_3)$		$f(x_1, x_2, x_3, x_4)$	
$x(3)$		$f(x_2, x_3, x_4)$		
	$f(x_3, x_4)$			
$x(4)$				

In the same way that the $(n+1)$st difference of an nth-order polynomial was zero, it is found that the nth-order divided difference of an nth-order polynomial is zero. Newton's interpolation formula, based on divided differences, is of the form

$$y = f(x_0) + (x - x_0)f(x_0, x_1) + (x - x_0)(x - x_1)f(x_0, x_1, x_2) + \cdots$$

$$+ (x - x_0)(x - x_1) \cdots (x - x_{n-1})f(x_0, x_1, \ldots, x_n) + \epsilon(x)$$

where

$$\epsilon(x) = \frac{f^{(n)}(\theta)}{n!} \prod_{k=0}^{n} (x - x_k),$$

θ is between the largest and smallest of $x, x_0, x_1, \ldots, x_n$

2-9 INVERSE INTERPOLATION

We have been concerned with the interpolation to determine values of the dependent variable, given either equally spaced or unequally spaced values of the independent variable. Inverse interpolation involves finding values of the independent variable, given a table of values of the dependent variable. In particular, the method is useful for finding missing values of the independent variable in a tabulated set of data. A nice feature of numerical analyses using finite or divided difference tables is that inverse interpolation is performed in identically the same way in which interpolation is conducted. That is, the procedure is an interpolation process where the dependent and independent variables are switched. The values of x are to be determined and thus are "dependent" on the values of y. Hence the interpolation problem is one of developing an interpolating polynomial through the sequence of values of the independent variables in the problem. The procedure is then identical to regular interpolation.

2-10 THIRTEEN-PLACE PRECISION FROM TWO-DIGIT TABLES

An interesting aspect of the use of difference tables that is consistent with the high precision of pocket calculators is that specially prepared difference tables permit precision interpolation to the accuracy of the calculator's capability, but with table entries of apparently only two or three significant digits. In fact, the accuracy is known to an *infinite* number of digits but only two significant digits are nonzero. For example, a table of $y = x^2$ can take either the form shown in Table 2-13 or that shown in Table 2-14.

Table 2-13   Difference Table for $y = x^2$ where $\Delta x = 1$, $x_0 = \pi$ (Two-Place Accuracy)[a]

x	y	Δy	$\Delta^2 y$	$\Delta^3 y$
3.14	9.87			
		7.27		
			2.01	
4.14	17.14			-0.01
		9.28		
5.14	26.42		2.00	
		11.28		0
6.14	37.70		2.00	
		13.28		0
7.14	50.98		2.00	
		15.28		
8.14	66.26			

[a] Interpolation formulas based on this difference table can only be accurate to two places after the decimal point at best, no matter how high the order of the interpolation formula, because the differences are only known to two places.

Both difference tables are developed with integer differences in the dependent variable and precisely known values of the independent variable in Table 2-14 and two-place accuracy in Table 2-13. The difference between these two tables is that the first permits interpolation to an accuracy of only two places, while the second interpolation permits an accuracy to 10 places, even though both are based on entries in the table that are only known to a few significant figures. It is precisely in this manner that the scientific pocket calculator, and even the simple four-function pocket calculator, can be used to boot-strap itself to generate advanced mathematical functions to an extremely high precision. All that is required is that certain values of both the dependent and independent variables of the advanced mathematical function be known precisely— where only a small number of nonzero digits make up the number. These values can then be used in a high-order difference table to generate an interpolation formula that will be very accurate over the range of the data table.

Table 2-14   Difference Table for $y = x^2$ where $\Delta x = 1$, $x_0 = 3$ (∞ Place Accuracy)[a]

x	y	Δy	$\Delta^2 y$	$\Delta^3 y$
3	9			
		7		
4	16		2	
		9		0
5	25		2	
		11		0
6	36		2	
		13		0
7	49		2	
		15		0
8	64		2	
		17		
9	81			

[a] Interpolation formulas based on this difference table can be as accurate as the order of the interpolation formula will allow *because the differences are known precisely.*

Example 2-1   Using the definition

$$\Delta = z - 1$$

write an expression that will interpolate between data points and differences in a data table. Since

$$z = (1 + \Delta)$$

$$z^n = (1 + \Delta)^n$$

then

$$z^n[y(x)] = y(x + n\Delta x) = y(x) + n\Delta y(x) + \frac{n(n-1)}{z}\Delta^2 y(x) + \cdots \Delta^n y(x)$$

Note that this is Newton's forward difference interpolation formula.

Example 2-2   Use a difference table to check an interpolating polynomial.

We can check Newton's or any other interpolation formula by substituting data points from a known polynomial such as $y = x^2$ (see Table 2-15) into the formula. Then $y(x + n\Delta x)$ becomes

$$y(0 + n) = y(n) = 0 + n + n(n - 1) = n + n^2 - n = y(n) = n^2$$

We see that the interpolation formula gives the original polynomial $y(x)$ $= x^2$ again: a result to be expected, since the interpolation formula is itself a polynomial.

Table 2-15 Numerical Example for $y = x^2$

x	y	Δy	$\Delta^2 y$	$\Delta^3 y$
0	0			
		1		
1	1		2	
		3		0
2	4		3	
		5		
3	9			

2-11 REFERENCES

For this chapter consult Richard Hamming's *Numerical Methods for Scientists and Engineers* (McGraw-Hill, 1973), Chapters 9 and 10.

PART TWO

NUMERICAL EVALUATION
OF FUNCTIONS
ON THE POCKET CALCULATOR

CHAPTER 3

ELEMENTARY ANALYSIS
WITH THE POCKET CALCULATOR

3-1 INTRODUCTION

A number of analytical topics used in elementary analysis are discussed here. Among them commonly used progressions including arithmetic, geometric, harmonic, and concepts of generalized means; the detailed definitions of absolute and relative error; nested parenthetical forms of commonly used infinite series including Taylor's series; certain often-encountered forms of the binomial series; the reversion of series; and methods for transforming series that converge slowly into series that converge more quickly. Also discussed are methods for evaluating the roots of polynomials including quadratics, cubics, quartics, and quintics; methods for the numerical evaluation of transcendental functions and for solving plane and spherical triangles; and methods for numerically evaluating commonly encountered functions of complex variables. The formulas and equations used for pocket calculator analysis are written in forms most convenient for evaluation on the pocket calculator.

3-2 NUMERICAL EVALUATION OF PROGRESSIONS

An arithmetic progression is defined by a sequence of numbers

$$a_n = a_1 + (n-1)d, \qquad (n \text{ an integer } > 0)$$

where a and d are real numbers. For $a_1 = 3e$ and $d = -\pi$

n	a_n
1	8.154845484
2	5.013252830
3	1.871660176
4	-1.269932478
5	-4.411525132

A common problem is to compute the sum of the arithmetic progression to n terms:

$$S_n(d) = a + (a+d) + (a+2d) + \cdots + [a + (n-1)d]$$

There are two formulas for computing the sum of an arithmetic progression. The first is

$$S_n(d) = na + \tfrac{1}{2}n(n-1)d$$

which can be rewritten in nested parenthetical form for easy evaluation on the pocket calculator as

$$S_n(d) = n\left(a + \frac{d}{2}(n-1)\right)$$

Another formula for computing the sum of the arithmetic progression to n terms is

$$S_n(d) = \frac{n}{2}(a+l)$$

Here the last term in the series l is

$$l = a + (n-1)d$$

We note that this equation is already in a form that can be easily evaluated on the pocket calculator.

The geometric progression is defined by a sequence of terms of the form

$$a_n = a_1 r^{n-1}, \qquad (n \text{ an integer} > 0)$$

where a and r are real numbers. For $a_1 = 3e$ and $r = -\pi$

n	a_n
1	8.154845484
2	-2.561920267×10^1
3	$+8.048509891 \times 10^1$
4	-2.528513955×10^2
5	$+7.943560867 \times 10^2$

The sum of the geometric progression to n terms is

$$S_n = a_1 + a_1 r + a_1 r^2 + a_1 r^3 + a_1 r^4 + \cdots + a_1 r^{n-1}$$

It can be computed with the formula

$$S_n = \frac{a_1(1 - r^n)}{1 - r} = \frac{a_1 - rl}{1 - r}$$

where l is the last term. If $r < 1$ in size, then as $n \to \infty$

$$\lim_{n \to \infty} (S_n) = \frac{a_1}{1 - r}$$

since the last term $l \to 0$. The sum of the geometric progression to n terms requires scratch-pad or memory storage. Table 3-1 shows a typical key stroke sequence needed for its evaluation and the required storage.

Three types of means are encountered in advanced analysis—the arithmetic mean, the geometric mean, and the harmonic mean. Though they are all special cases of the generalized mean

$$M(t) = \left(\frac{1}{n} \sum_{k=1}^{n} a_k^t \right)^{1/t}$$

we are explicit here and write them out. The arithmetic mean of n quantities is defined by the equation

$$A_n = \frac{a_1 + a_2 + a_3 + \cdots + a_n}{n}$$

which can be computed conveniently (though not so easily as summing and dividing by the total number of samples) on the pocket calculator

**Table 3-1   Typical Key Stroke Sequences for Evaluating
the Sum of Terms in a Geometric Progression**

Algebraic			Reverse-Polish	
(r)	$+$		(r)	(1.0)
y^x	(1.0)		$\uparrow$	$+$
(n)	$1/x$		(n)	$\div$
$=$	$\times$		y^x	$\boxed{S_n}$
CHS	RCL		CHS	
$+$	$=$		(1.0)	*Note*: Recall is
(1.0)	$\boxed{S_n}$		$+$	automatic in
(a_1)			$\times$	Reverse-Polish
STO			(r)	for stack memory
(r)			CHS	(HP-35&21) but not
CHS			$\uparrow$	for register
				memory (HP-45,55&65)

$(\ \ )\rightarrow$data entry.
$\square\rightarrow$output.

using a recursion formula

$$A_{n+1} = \frac{1}{n+1}(nA_n + a_{n+1})$$

which can be developed from the equation for A as follows:

$$A_n = \frac{1}{n}\sum a_i$$

$$A_{n+1} = \frac{1}{n+1}\left[\sum a_i + a_{n+1}\right]$$

$$\left(\frac{n+1}{n}\right)A_{n+1} = \frac{\sum a_i}{n} + \frac{a_{n+1}}{n} = A_n + \frac{a_{n+1}}{n}$$

$$.A_{n+1} = \frac{n}{n+1}A_n + \frac{1}{n+1}a_{n+1} = \frac{1}{n+1}[nA_n + a_{n+1}]$$

An advantage to using a recursive "averager" is that the analyst can observe the convergence of the mean as he adds more terms to the calculation. He can thus often reduce the workload in computing an average by using only the numbers that are necessary to estimate the mean to the accuracy he desires.

The recursive form is directly implementable, using the key strokes shown in Table 3-2.

Table 3-2 Typical Key Stroke Sequences for Recursive Arithmatic Averaging

Algebraic	Reverse-Polish

( )→ data entry.
□→ output
◯→ mental step done by analyst.
*Initial conditions can be $A_n = 0$ when $n = 0$.

The geometric mean of n quantities is defined by the relationship

$$G = (a_1 a_2 \cdots a_n)^{1/n}, \qquad (a_i > 0,\, i = 1, 2, \ldots, n)$$

which is easily calculated using the recursion formula for the geometric mean of n quantities as given by

$$G_{n+1} = (a_{n+1} G_n^n)^{1/n+1}$$

and is developed as follows:

$$G_n = \left(\prod^n a_i \right)^{1/n}$$

$$G_n^n = \prod^n a_i$$

$$G_{n+1}^{n+1} = \left(\prod^n a_i \right)(a_{n+1})$$

$$G_{n+1}^{n+1/n} = (a_{n+1})^{1/n} \left(\prod^n a_i \right)^{1/n} = a_{n+1}^{1/n} G_n$$

$$G_{n+1} = a_{n+1}^{1/n+1} G_n^{n/n+1} = (a_n G_n^n)^{1/n+1}$$

A typical sequence of key strokes for evaluating this equation appears in Table 3-3.

Table 3-3 Typical Key Stroke Sequence for Recursive Evaluation of the Geometric Mean

Algebraic	Reverse-Polish
G_n	G_n
y^x	$\uparrow$
(n)	(n)
$\times$	y^x
(a_{n+1})	(a_n)
y^x $\;\;n = n+1$	$\times$ $\;\;n = n+1$
$(n+1)$	$\uparrow$
$1/x$	$(n+1)$
$=$	$1/x$
$\boxed{G_{n+1}}$	y^x
	$\boxed{G_{n+1}}$

( )→ data input.
□→ output.
◯ → mental step by analyst.

The harmonic mean of n quantities is defined by

$$\frac{1}{H} = \frac{1}{n}\left(\frac{1}{a_1} + \frac{1}{a_2} + \cdots + \frac{1}{a_n}\right), \qquad (a_i > 0, i = 1, 2, \ldots, n)$$

It, too, can be evaluated by using a recursion formula:

$$H_{n+1} = \left\{\frac{1}{n+1}\left(\frac{1}{a_{n+1}} + \frac{n}{H_n}\right)\right\}^{-1}$$

The harmonic mean is evaluated using the typical key stroke sequence given in Table 3-4.

Finally, the generalized mean is related to the geometric, arithmetic, and

Table 3-4 Typical Key Stroke Sequence for Recursive Evaluation of the Harmonic Mean

Algebraic	Reverse-Polish

H_n
$1/x$
$\times$
(n)
STO
(a_{n+1})
$1/x$
$+$
RCL
STO
$(n+1)$
$\times$
RCL
$1/x$
$\boxed{H_{n+1}}$

$n = n+1$

H_n
$1/x$
(n)
$\times$
(a_{n+1})
$1/x$
$+$
$(n+1)$
$1/x$
$\times$
$1/x$
$\boxed{H_{n+1}}$

$n = n+1$

( )→ data input.
□→ output.
◯ → mental step by analyst.

harmonic means according to the relations

$$\lim_{t \to 0} M(t) = G$$

$$M(1) = A$$

$$M(-1) = H$$

3-3 THE DEFINITION OF ABSOLUTE AND RELATIVE ERROR

We discussed absolute and relative errors previously in the context of other matters. In the next chapter a number of errors are quoted; hence it is important to define precisely what is meant by absolute and relative errors. When x_0 is an approximation to the true value of x, we say the following:

1. The absolute error of x_0 is $\Delta x = x_0 - x = (\text{calculated} - \text{true})$.

2. The relative error of x_0 is $\delta x = \Delta x / x$, (calculated $-$ true)/true, which is approximately equal to $\Delta x / x_0$.

3. The percentage error is 100 times the relative error.

If in (2) we use the approximation of the true value of x to estimate percentage error then in a sense there is a small error in estimating the relative error.

The absolute error of the sum or difference of several numbers is *at most* equal to the sum of the absolute errors of the individual numbers. If it can be assumed that the errors occur in a random independent fashion, a more reasonable estimate of the error in computing the sum or difference of several numbers is root-sum-square error defined as

$$\left(\sum \Delta x_i^2 \right)^{1/2}$$

The relative error of the product or quotient of several factors is at most equal to the sum of the relative errors of the individual factors. Finally, if $y = f(x)$, the relative error

$$\delta y = \frac{\Delta y}{y} \cong \frac{f'(x)}{f(x)} \Delta x$$

If we have

$$y = f(x_1, x_2, \ldots, x_n)$$

and the absolute error in x_i is Δx_i for all n, then the absolute error in f is

$$\Delta f \approx \frac{\partial f}{\partial x_1} \Delta x_1 + \frac{\partial f}{\partial x_2} \Delta x_2 + \cdots + \frac{\partial f}{\partial x_n} \Delta x_n$$

Simple rules, similar to those for the relative error of a product or the quotient, can easily be derived for relative errors of powers and roots. It turns out that the relative error of an nth power is almost exactly n times the relative error of the base power, while the relative error of an nth root is $1/n$th of the relative error of the radicand.

Calculations with Approximate Values

Where they are developed from test experiments or from tables of characteristics of physical systems, data are usually inaccurate to some degree. In general, calculations made with data based on measurements involve errors of some magnitude. Another type of error in calculating with approximate values is due to the use of numerical values of numbers that are truncated, producing roundoff errors. The maximum errors associated with these effects can be estimated. When the rounding is done correctly, the roundoff is at most one-half of the unit in the last place retained in the number.

When the numbers are rounded, the addition of zeros after the last digit of the decimal fraction makes a difference. The number 0.98700 is stated with 100 times greater accuracy than 0.987. In the first case the number implies that at most its error is 5×10^{-6}. In the second case, the error can be as large as 5×10^{-4}. The implications of accuracy should be stated precisely when tabulating results computed on the pocket calculator.

The error due to a calculation that results from an inaccuracy of the data is known historically as "error of data." The error introduced into the calculation by way of approximation associated with the limitations of the machine or field of numbers being used in the calculation is historically called "error of calculation." It is the objective of any calculation to make the error of calculation significantly less than the error of data. Fortunately, for most pocket calculators the size of the numbers that can be contained is so large that the error of engineering calculation is almost always substantially smaller than the error of engineering data.

When good computing practice is followed, care must be taken when computing the difference between nearby numbers. Since occasionally the magnitude of the error of calculation is found to determine the method of the calculations to be done, we are interested in estimating from the error of data the maximum error to be expected in the result of the calculation due to this error in the data. It is for these reasons that we give the rules for computing the absolute errors of the sums, differences, products, and quotients of numerical calculations. These formulas can be used to answer the questions about the size of the error-of-data from which can be determined whether the error-of-calculation will be on the same order of magnitude or smaller. Furthermore, if the error-of-calculation is less, it can be used to guide the analyst in how much the error-of-data will limit the accuracy of a calculation. This will indicate the accuracy remaining after a complex or involved calculation.

The results of a calculation are the most inaccurate when the difference of two nearly equal and only approximately known numbers are involved. To determine the relative error in these cases, the sum of the absolute errors, taken without regard to sign, is divided by the difference of the two numbers involved (a small number that can turn even a small absolute error into a large relative error).

3-4 INFINITE SERIES

We will have many occasions to use Taylor's formula for a single variable as given by the expression

$$f(x+h) = f(x) + hf'(x) + \frac{h^2}{2}f''(x) + \cdots + \frac{h^{n-1}}{(n-1)!}f^{n-1} + Rn$$

This equation has an error formula that can be written in three typical forms:

$$R_n = \frac{h^n}{n!}f^n(x+\theta_1 h), \qquad (0<\theta_1<1)$$

$$R_n = \frac{h^n}{(n-1)!}(1-\theta_2)^{n-1}f^n(x+\theta_2 h), \qquad (0<\theta_2<1)$$

$$R_n = \frac{h^n}{(n-1)!}\int_0^1 (1-t)^{n-1}f^n(x+th)\,dt$$

The truncated version of the series can be expanded in nested parenthetical forms for convenient numerical evaluations when the numerical values of the derivative either are given or can be quickly computed.

$$f_1 = f(x)$$

$$f_2 = f(x) + hf'(x)$$

$$f_3 = f(x) + h\left(f' + \frac{hf''}{2}\right)$$

$$f_4 = f(x) + h\left(f' + \frac{h}{2}\left(f'' + \frac{hf'''}{3}\right)\right)$$

$$f_5 = f(x) + h\left(f' + \frac{h}{2}\left(f'' + \frac{h}{3}\left(f''' + \frac{hf''''}{4}\right)\right)\right)$$

$$\vdots$$

$$f_n = f(x) + h\left(f' + \frac{h}{2}\left(f'' + \frac{h}{3}\left(f''' + \frac{h}{4}\left(f'''' + \cdots\right.\right.\right.\right.$$

$$\left.\left.\left.\left. + \frac{h}{n-1}\left(f^{n-1} + \frac{hf^n}{n}\right)\right)\cdots\right)\right)\right)$$

Taylor series expansions of $f(x)$ on the point a are given by the expression

$$f(x)=f(a)+(x-a)f'(a)+\frac{(x-a)^2}{2}f''(a)+\cdots+\frac{(x-a)^{n-1}}{(n-1)!}f^{n-1}(a)+R_n$$

where the remainder formula is given by

$$R_n=\frac{(x-a)^n}{n!}f^n(\zeta),\qquad (a<\zeta<x)$$

This expression, too, can be written in nested parenthetical form as

$$f_n=f(a)+(x-a)\left(f'(a)+\frac{(x-a)}{2}\left(f''(a)+\frac{(x-a)}{3}\left(f'''(a)+\cdots\right.\right.\right.$$

$$\left.\left.\left.+\frac{(x-a)}{n-1}\left(f^{n-1}(a)+\frac{(x-a)}{n}f^n(a)\right)\right)\cdots\right)\right)$$

Binomial Series

The binomial series is encountered many times in combinatorial analysis as well as in the formulation of difference equations for numerical analysis. The binomial series can be written in the general form

$$(1+x)^\alpha=\sum_{k=0}^{\alpha}\binom{\alpha}{k}x^k,\qquad (-1<x<1)$$

where

$$\binom{\alpha}{k}=\frac{\alpha!}{(\alpha-k)!k!}$$

Particular series of interest are

$$(1+x)^\alpha=1+\alpha x+\frac{\alpha(\alpha-1)}{2!}x^2+\frac{\alpha(\alpha-1)(\alpha-2)}{3!}x^3+\cdots$$

which can be written in nested parenthetical form for easy pocket calculator evaluation as

$$(1+x)^{\alpha} = 1 + \alpha x \left(1 + \frac{x(\alpha-1)}{2} \left(1 + \frac{x(\alpha-2)}{3} \left(1 + \frac{x(\alpha-3)}{4} + \cdots \right. \right. \right.$$

$$\left. \left. \left. + \frac{x(\alpha-n+1)}{n-1} \left(1 + \frac{x(\alpha-n)}{n} \right) \cdots \right) \right) \right)$$

Other frequently encountered binomial series are the following:

(a)

$$(1+x)^{-1} = 1 - x + x^2 - x^3 + x^4 - x^5 + \cdots, \qquad (|x|<1)$$

$$\therefore (1+x)^{-1} = 1 - x\big(1 - x\big(1 - x\big(1 - x(\cdots,)\big)\big)\big) \qquad (|x|<1)$$

(b)

$$(1+x)^{1/2} = 1 + \frac{x}{2} - \frac{x^2}{8} + \frac{x^3}{16} - \frac{5x^4}{128} + \frac{7x^5}{256} - \frac{21x^6}{1024}, \qquad (|x|<1)$$

$$\therefore (1+x)^{1/2} = 1 + \frac{x}{2} \left(1 - \frac{x}{4} \left(1 - \frac{x}{2} \left(1 - \frac{5}{8} x(\cdots,) \right) \right) \right) \qquad (|x|<1)$$

(c)

$$(1+x)^{-1/2} = 1 - \frac{x}{2} \left(1 - \frac{3x}{4} \left(1 - \frac{5x}{6} \left(1 - \frac{7x}{8} (1\cdots,) \right) \right) \right) \qquad (|x|<1)$$

(d)

$$(1+x)^{1/3} = 1 + \frac{x}{3} \left(1 - \frac{x}{3} \left(1 - \frac{5x}{9} \left(1 - \frac{2x}{6} (1\cdots,) \right) \right) \right) \qquad (|x|<1)$$

(e)

$$(1+x)^{-1/3} = 1 - \frac{x}{3} \left(1 - \frac{2x}{3} \left(1 - \frac{7x}{9} \left(1 - \frac{5x}{6} (1\cdots,) \right) \right) \right) \qquad (|x|<1)$$

Operations with Truncated Forms of Infinite Series

An integral part of advanced analysis on the pocket calculator is the numerical evaluation of truncated series. Generally, the approach is to truncate the series at something on the order of four terms and use the series to evaluate the function over the region that has a good fit with the function being considered. Once a series is generated, whether with Chebyshev polynomials, Taylor series, the binomial series, Legendre polynomials, or some other means, such operations can be performed on the series as inverting the series, taking the square root of it, squaring it, multiplying or dividing it, taking the exponential of it, or taking the logarithm of it. This is conveniently done by manipulating the coefficients in the series. These operations are tabulated in Table 3-5 for the three series

$$s_1 = 1 + a_1 x + a_2 x^2 + a_3 x^3 + a_4 x^4 + \cdots$$

$$s_2 = 1 + b_1 x + b_2 x^2 + b_3 x^3 + b_4 x^4 + \cdots$$

$$s_3 = 1 + c_1 x + c_2 x^2 + c_3 x^3 + c_4 x^4 + \cdots$$

Among convenient series manipulations is the reversion of series, where the dependent variable is solved in terms of the independent variable. Given the series

$$y = ax + bx^2 + cx^3 + dx^4 + ex^5 + fx^6$$

we can write x as a function of y as

$$x \approx Ay + By^2 + Cy^3 + Dy^4 + Ey^5 + Fy^6$$

where

$$A = \frac{1}{a}$$

$$B = -\frac{b}{a^3}$$

$$C = \frac{2b^2 - ac}{a^5}$$

$$D = \frac{5abc - a^2 d - 5b^3}{a^7}$$

$$E = \frac{6a^2 bd + 3a^2 c^2 + 14b^4 - 21ab^2 c - a^3 e}{a^9}$$

$$F = \frac{7a^3 be + 7a^3 cd + 84ab^3 c - a^4 f - 28a^2 bc^2 - 42b^5 - 28a^2 b^2 d}{a^{11}}$$

Table 3-5 Series Operations

Operation	c_1	c_2	c_3	c_4
$s_3 = s_1^n$	na_1	$\frac{1}{2}(n-1)c_1a_1 + na_2$	$c_1a_2(n-1) + \frac{c_1a_1^2}{6}(n-1)(n-2) + na_3$	$na_4 + c_1a_3(n-1) + \frac{1}{2}n(n-1)a_2^2 + \frac{1}{2}(n-1)(n-2)c_1a_1a_2 + \frac{1}{24}(n-1)(n-2)(n-3)c_1a_1^3$
$s_3 = s_1s_2$	$a_1 + b_1$	$b_2 + a_1b_1 + a_2$	$b_3 + a_1b_2 + a_2b_1 + a_3$	$b_4 + a_1b_3 + a_2b_2 + a_3b_1 + a_4$
$s_3 = s_1/s_2$	$a_1 - b_1$	$a_2 - (b_1c_1 + b_2)$	$a_3 - (b_1c_2 + b_2c_1 + b_3)$	$a_4 - (b_1c_2 + b_2c_2 + b_3c_1 + b_4)$
$s_3 = e^{(s_1-1)}$	a_1	$a_2 + \frac{a_1^2}{2}$	$a_3 + a_1a_2 + \frac{a_1^3}{6}$	$a_4 + a_1a_3 + \frac{a_2^2}{2} + \frac{a_2a_1^2}{2} + \frac{a_1^4}{24}$
$s_3 = 1 + \ln(s_1)$	a_1	$a_2 - \frac{a_1c_1}{2}$	$a_3 - \frac{(a_2c_1 + 2a_1c_2)}{2}$	$a_4 - \frac{(a_3c_1 + 2a_2c_2 + 3a_1c_3)}{4}$

Transformation of Series

Occasionally, slow-converging series are encountered in numerical analysis where the object is to compute the sum of the series to high accuracy. Usually we would use some form of economization to improve the accuracy of such a series (see Chapter 8). We may, however, also know another series that can be used to improve the convergence (accuracy) of the original series. This is convenient when numerically evaluating the sum of a slowly converging series of the form

$$s = \sum_{k=0}^{\infty} a_k$$

where it is known that the series does in fact converge and where we have another series

$$c = \sum_{k=0}^{\infty} c_k$$

which is also convergent and which we know to have the sum c and the limit of a_k / c_k as k approaches infinity to equal λ (where λ is not equal to zero); then

$$s = \lambda c + \sum_{k=0}^{\infty} \left(1 - \lambda \frac{c_k}{aK}\right) a_k$$

This technique is known as Kummer's transformation. It transforms one series into another that is more convenient for numerical evaluation. While not developed originally for this purpose, it turns out to be quite useful in numerical evaluation of slowly converging series.

   Another approach to numerically evaluating a truncated series is to use the Euler-Maclaurin summation formula. This is another technique for numerically evaluating series using another series that converges more quickly. Provided that the difference of derivatives at the end points of the interval over which the series is being evaluated is small, the Euler-Maclaurin summation formula is

$$s = \sum_{k=1}^{n-1} f_k \cong \int_0^n f(k)\,dk - \tfrac{1}{2}(f_0 - f_n) + \tfrac{1}{12}\left(f_n^{(\mathrm{I})} - f_0^{(\mathrm{I})}\right)$$

$$- \tfrac{1}{720}\left(f_n^{(\mathrm{III})} - f_0^{(\mathrm{III})}\right) + \frac{\left(f_n^{(\mathrm{V})} - f_0^{(\mathrm{V})}\right)}{30240}$$

3-5 THE SOLUTION OF POLYNOMIALS

The numerical solution of a polynomial on the pocket calculator involves a clear understanding of the possible location of the polynomial's roots in the complex plane. For this reason, we take a few moments to refresh our understanding of algebraic equations. It should be remembered that an nth-order algebraic equation has n roots. If the coefficients in the polynomial are real, the roots of the equation are either all real, some being equal and some not, or have pairs of roots that are complex conjugates of each other and other roots that are real with various locations on the real axis. The occurrence of complex roots in complex conjugate pairs arises from our assumption that the coefficients in the polynomials are real, not complex. If the coefficients are complex, of course, the roots can occur anywhere in the complex plane. In this book we concern ourselves only with polynomials that have real coefficients, since they are the most frequently encountered algebraic equations in engineering analysis.

The Solution of Quadratic Equations

If we are given a quadratic equation of the form

$$az^2 + bz + c = 0$$

its roots can be numerically evaluated with the formula

$$z_1 = -\left(\frac{b}{2a}\right) + \frac{\sqrt{q}}{2a}$$

$$z_2 = -\left(\frac{b}{2a}\right) - \frac{\sqrt{q}}{2a}$$

where

$$q = b^2 - 4ac$$

From time to time we will make use of the following easily verified properties of the roots:

$$z_1 + z_2 = -\frac{b}{a}$$

$$z_1 z_2 = \frac{c}{a}$$

It is apparent from the equations for the two roots that

1. If $q > 0$, the two roots will be real and unequal.
2. If $q = 0$, the two roots are both real and equal.

3. If $q<0$, the roots occur in complex conjugate pairs.

The numerical evaluation on the pocket calculator should involve first the calculation to determine q and then the use of equations for the roots for their evaluation once the situation of the roots is determined.

Solution of Cubic Equations

If we are given a cubic equation of the form

$$z^3 + a_2 z^2 + a_1 z + a_0 = 0$$

the first step in computing its roots is to calculate q and r:

$$q = \frac{a_1}{3} - \frac{a_2^2}{9}$$

$$r = \frac{a_1 a_2 - 3 a_0}{6} - \frac{a_2^3}{27}$$

Then:

1. If $q^3 + r^2 > 0$, the cubic equation has one real root and a pair of complex conjugate roots.
2. If $q^3 + r^2 = 0$, all the roots are real and at least two are equal.
3. If $q^3 + r^2 < 0$, all roots are real and unequal (the irreducible case).

Once the nature of the roots is known, it is a simple matter to use the following equations to evaluate the roots on the pocket calculator. First, compute

$$s_1 = \left[r + (q^3 + r^2)^{1/2} \right]^{1/2}$$

$$s_2 = \left[r - (q^3 + r^2)^{1/2} \right]^{1/2}$$

Then the roots can be calculated from an understanding of their nature and the following three equations:

$$z_1 = (s_1 + s_2) - \frac{a_2}{3}$$

$$z_2 = \frac{-(s_1 + s_2)}{2} - \frac{a_2}{3} + \frac{i\sqrt{3}}{2}(s_1 - s_2)$$

$$z_3 = -\frac{(s_1 + s_2)}{2} - \frac{a_2}{3} + \frac{i\sqrt{3}}{2}(s_1 - s_2)$$

Note that if $q^3 + r^2 = 0$, s_1 will equal s_2 and the imaginary component of the roots will drop out, leaving the two z roots, z_2 and z_3, equal, while z_1 may not necessarily be equal, depending on the value of s_2.

Once the roots of the cubic equation are evaluated, they satisfy the following relations:

$$z_1 + z_2 + z_3 = -a_2$$

$$z_1 z_2 + z_1 z_3 + z_2 z_3 = a_1$$

$$z_1 z_2 z_3 = -a_0$$

These relations can be used as a check on the calculation of the roots.

The process of numerically evaluating the roots of the quartic equation is somewhat involved, even for pocket calculator evaluation. Under some conditions, however, simple evaluations can be made. For example, consider the quartic equation

$$z^4 + a_3 z^3 + a_2 z^2 + a_1 z + a_0 = 0$$

One approach to evaluating the roots of this quartic equation is to find the real root of the cubic equation

$$\mu^3 - a_2 \mu^2 + (a_1 a_3 - 4a_0)\mu - (a_1^2 + a_0 a_3^2 - 4a_0 a_2) = 0$$

and then determine the four roots, of the quartic equation as solutions to the two quadratic equations

$$v^2 \left[\frac{a_3}{2} - \left(\frac{a_3^2}{4} + \mu_1 - a_2 \right)^{1/2} \right] v + \frac{\mu_1}{2} - \left[\left(\frac{\mu_1}{2} \right)^2 - a_0 \right]^{1/2} = 0$$

$$v^2 + \left[\frac{a_3}{2} + \left(\frac{a_3^2}{4} + \mu_1 - a_2 \right)^{1/2} \right] v + \frac{\mu_1}{2} + \left[\left(\frac{\mu_1}{2} \right)^2 - a_0 \right]^{1/2} = 0$$

Once the roots of the quartic are evaluated and can be written in the form

$$z^4 + a_3 z^3 + a_2 z^2 + a_1 z + a_0 = (z^2 + p_1 z + q_1)(z^2 + p_2 z + q_2)$$

the following conditions hold:

$$p_1 + p_2 = a_3$$

$$p_1 p_2 + q_1 + q_2 = a_2$$

$$p_1 q_2 + p_2 q_1 = a_1$$

$$q_1 q_2 = a_0$$

Finally, if z_1, z_2, z_3, z_4 are the roots of the quartic equation, the following conditions hold among the roots:

$$z_1 + z_2 + z_3 + z_4 = -a_3$$

$$\sum z_i z_j z_k = -a_1$$

$$\sum z_i z_j = a_2$$

$$z_1 z_2 z_3 z_4 = a_0$$

Again, these conditions can be used to check on the calculation of the roots.

The evaluation of the roots of a polynomial up to quartics is tedious and usually inaccurate (at best) on a slide rule, by hand analysis, or even on the old mechanical calculators (though accurate); it is a relatively fast and accurate process on the pocket calculator, however.

3-6 SUCCESSIVE APPROXIMATION METHODS

Again, we are concerned with the problem of determining the roots of an equation, but the equation is of a more general form. We are looking for the condition

$$f(x) = 0$$

That is, we are looking for the values of x such that $f(x)$ will equal zero. In this case $f(x)$ need not be a polynomial in x. If we let $x = x_n$, the approximation of the root, then when f_n is not equal to 0 it is equal to ϵ (the error). If we now use ϵ to update our estimate of the root,

$$\Delta x = c_n \epsilon_n = c_n f_n$$

we can write

$$x_{n+1} = x_n + c_n f(x_n), \qquad (n = 1, 2, 3 \cdots) \tag{3-1}$$

When it is found that $f'(x)$ is greater than or equal to zero and the constants c_n are negative and bounded, the sequence of x_n converges monotonically to the root $x = r$. If c is a constant less than zero and f' is greater than zero, the process converges but not necessarily monotonically. A number of approaches have been developed to compute c_n. Among these are the *regula falsi* method, the method of successive iterations, Newton's method, and the Newton-Raphson method. The *regula falsi* method begins with the assumption that we are given $y = f(x)$; the objective is to find $x = r$ such that $f(r) = 0$. We choose a pair of values of x, x_0, and x_1 such that $f(x_0)$ and $f(x_1)$ have opposite signs. Then equation 3-1 can take the form

$$x_2 = x_1 - \left(\frac{x_1 - x_0}{f_1 - f_0} \right) f_1 = \frac{f_1 x_0 - f_0 x_1}{f_1 - f_0} \tag{3-2}$$

The third- and higher-order estimates of the root x_n are computed using x_2 and either x_0 or x_1 for which $f(x_0)$ or $f(x_1)$ is of opposite sign to $f(x_2)$.

This method is equivalent to an inverse interpolation. This is apparent from the form of equation 3-2.

In the method of successive iterations, the approach is to write the equation in an implicit form and use successive iterations to solve the equation $x = F(x)$. The iteration scheme is to compute

$$x_{n+1} = f(x_n)$$

The sequence of solutions to this implicit equation will converge to a zero of $x = F(x)$ if there exists a q such that

$$|f'(x)| \leqslant q < 1 \qquad \text{for} \quad a \leqslant x \leqslant b$$

and

$$a \leqslant x_0 \pm \frac{|f(x_0) - x_0|}{1 - q} \leqslant b$$

This is an attractive method for use on the pocket calculator because it does not involve remembering special formulas such as those associated with the *regula falsi* or the Newton (Newton-Raphson) methods. The problem encountered in applying the method of successive iterations on the implicit form of the equation whose roots are to be determined is that the implicit equation may not converge as quickly as other methods based on additional information (such as the derivatives of $f(x)$) whose function it is to ensure rapid convergence of the method.

Newton's method is to compute recursively estimates of the roots of the function $f(x)$ using the formula

$$x_{n+1} = x_n - \frac{f(x_n)}{f'(x_n)} \tag{3-3}$$

where $x = x_n$ is an approximation to the solution, $x + r$, of $f(x) = 0$. The sequence of solutions generated with Newton's rule will converge quadratically to $x = r$. The condition for monotonic convergence is that the product $f(x_0)f''(x_0)$ is greater than zero, and $f'(x)$ and $f''(x)$ do not change sign in the interval (x_0, r). The conditions for oscillatory convergence are also straightforward. When the product $x(x_0)f''(x_0)$ is less than zero and $f'(x)$ and $f''(x)$ do not change sign in the interval (x_0, x_1), equation 3-3 will converge, though it will oscillate. These conditions only hold, of course, when

$$x_0 \leqslant r \leqslant x_1$$

When Newton's method is applied to the evaluation of nth roots, we find that given $x^n = N$, if x_k is an approximation of $x = N^{1/n}$ then a sequence of improved x_k can be generated:

$$x_{k+1} = \frac{1}{n}\left(x_k\left(\frac{N}{x_k^n} + n - 1\right)\right)$$

This method will converge quadratically to x for all n, and is particularly useful for computing the nth roots iteratively on the four-function pocket calculator as covered in Chapter 1. It is derived here to show the procedure:

1. We wish to compute $x = (N)^{1/n}$.
2. Form $f(x) = (x^n - N) = 0$ from (1).
3. For Newton's rule, $x_{k+1} = x_k - f(x_k)/f'(x_k)$, we need $f(x_k)$ and $f'(x_k)$.
·4. $f(x_k) = (x_k^n - N)$ and $f'(x_k) = (nx_k^{n-1} - 0)$.
5. Substituting the results of (4) into (3) we find (6).
6.

$$x_{k+1} = x_k - \left(\frac{x_k^n - N}{nx_k^{n-1}}\right) = \frac{nx_k}{n} - \frac{x_k^n + N}{nx_k^{n-1}}$$

$$= \frac{1}{n}\left[\frac{N}{x_k^{n-1}} + (n-1)x_k\right] = \frac{1}{n}\left(x_k\left(\frac{N}{x_k^n} + n - 1\right)\right)$$

Details on finding the zeros of functions—an important subject in numerical analysis are given in Chapter 9.

3-7 ELEMENTARY TRANSCENDENTAL FUNCTIONS

In Chapter 1 we presented polynomial approximations for most of the transcendental functions found on the keyboard of the scientific calculator so that they could be evaluated on the simple four-function calculator. Not presented there, however, were approximations in terms of Chebyshev polynomials. Because the approximation in terms of Chebyshev polynomials is a mini-max approximation (minimizes the maximum error on the interval -1 to $+1$), they are accurate and useful, and for the sake of completeness they are presented here. The discussion of the reduction of order of series approximations to function in terms of Chebyshev polynomials is covered in Chapter 8 and their numerical evaluation is covered more fully in Chapter 4. For now we concern ourselves with the numerical evaluation of the elementary transcendental functions using Chebyshev polynomials.

Evaluating the natural log of y for y near zero can be difficult, at best. If y is near zero it is convenient to write

$$\ln(y)$$

in the form

$$\ln(1+x), \ y = 1 + x$$

Then we can write $\ln(1+x)$ as

$$\ln(1+x) = \sum_{n=0}^{\infty} A_n T_n(x), \qquad (0 \leqslant x \leqslant 1)$$

where the coefficients A_n are

n	A_n
0	0.376452813
1	0.343145750
2	-0.029437252
3	0.003367089
4	-0.000433276
5	0.000059471
6	-0.000008503
7	0.000001250
8	-0.000000188
9	0.000000029
10	-0.000000004
11	0.000000001

and where

$$T_n(x) = 2xT_{n-1}(x) - T_{(n-2)}(x)$$

$$T_0 = 1$$

$$T_1 = x$$

In a similar way, we can use Chebyshev polynomials to evaluate both

$$e^x \text{ and } e^{-x}$$

The coefficients for evaluating $e^{-x} = \sum\limits_{n=0}^{\infty} A_n T_n(x)$ are

n	A_n
0	0.64535270
1	-0.312841606
2	0.038704116
3	-0.003208683
4	0.000199919
5	-0.000009975
6	0.000000415
7	-0.000000015

The coefficients for evaluating e^x are

n	A_n
0	1.753387654
1	0.850391654
2	0.105208694
3	0.008722105
4	0.000543437
5	0.000027115
6	0.000001128
7	0.00000040
8	0.000000001

Again the restriction x is that

$$0 < x < 1$$

Now that we have a few example functions to work with, consider the procedure for using the Chebyshev polynomials to numerically evaluate these functions:

Step 1   Let the objective be to evaluate e^x near $x = x_0$, where x_0 is *not* on the interval $0 \leqslant x \leqslant 1$. Rewrite e^x so that the exponent is on the interval $[0, 1]$. For the sake of this discussion, we use

$$e^x = e^{(x - x_0) + x_0} = e^{x_0}[e^{(x - x_0)}] = e^{x_0} e^y$$

Then for x on the interval

$$x_0 \leqslant x \leqslant x_0 + 1$$

y is on the interval

$$0 \leqslant y \leqslant 1$$

Step 2   Select x and compute y.

Step 3   Compute $T_1 = y$.

Step 4   Compute $T_2 = 2yT_1 - 1$.
Compute $T_3 = 2yT_2 - T_1$.

$$
\begin{array}{ccccc}
\cdot & \cdot & \cdot & \cdot & \cdot \\
\cdot & \cdot & \cdot & \cdot & \cdot \\
\cdot & \cdot & \cdot & \cdot & \cdot \\
\end{array}
$$

Compute $T_n = 2yT_{n-1} - T_{n-2}$.

Step 5   Compute $e^y \cong \sum_0^n A_m T_m$ using the appropriate A_n.

Step 6   Compute $e^x = e^{x_0} e^y$.

Usually x_0 is chosen to be a convenient number for precise evaluation of e^{x_0} using the prime factors method (presented in Chapter 1). For example, if $x = 100$, then e^{100} becomes

$$\left(\left(\left(\left((e)^2\right)^2\right)^5\right)^5\right)^5$$

which is easily evaluated with a table lookup of e and 13 (at most) data entry key strokes plus 16 multiply key strokes on the four-function calculator.

The Chebyshev approximations for sine and cosine are given by the relation

$$\sin\left(\frac{\pi x}{2}\right) = x \sum_{n=0}^{\infty} A_n T_n(x^2)$$

and

$$\cos\left(\frac{\pi x}{2}\right) = \sum_{n=0}^{\infty} A_n T_n(x^2)$$

using the coefficients for A_n:

	Sine		Cosine
n	A_n	n	A_n
0	1.276278962	0	0.472001216
1	−0.285261569	1	−0.499403258
2	0.009118016	2	0.027992080
3	−0.000136587	3	−0.000596695
4	0.000001185	4	0.000006704
5	−0.000000007	5	−0.000000047

Here x must reside in the interval

$$|x| \leqslant 1$$

Formulas for the Solution of Plane and Spherical Triangles

Many elementary analysis problems involve the solution of triangles. These include plane right triangles, plane triangles, and spherical triangles. Consider the plane right triangle shown in Figure 3-1. Here A, B, and C are the vertices of the triangle and a, b and c are their opposite sides. Then

$$\sin A = \frac{a}{c} = \frac{1}{\csc A}$$

$$\cos A = \frac{b}{c} = \frac{1}{\sec A}$$

$$\tan A = \frac{a}{b} = \frac{1}{\cot A}$$

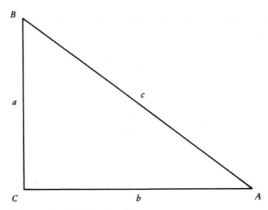

Figure 3-1   Right triangle.

Now consider Figure 3-2. This plane triangle has angles A, B, C and sides opposite a, b, c. The law of sines states that

$$\frac{a}{\sin A} = \frac{b}{\sin B} = \frac{c}{\sin C}$$

and the law of cosines is

$$\cos A = \frac{c^2 + b^2 - a^2}{2bc}$$

Also, the following four relationships hold for plane triangles:

$$a = b \cos C + c \cos B$$

$$\frac{a+b}{a-b} = \frac{\tan \frac{1}{2}(A+B)}{\tan \frac{1}{2}(A-B)}$$

$$\text{area} = \frac{bc \sin A}{2}$$

$$\text{area} = [s(s-a)(s-b)(s-c)]^{\frac{1}{2}}$$

where $s = \frac{1}{2}(a+b+c)$.

Figure 3-3 shows a spherical triangle with angles A, B, C and sides opposite a, b, c. The four commonly used formulas in spherical tri-

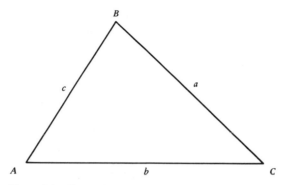

Figure 3-2 Plane triangle.

gonometry are

$$\frac{\sin A}{\sin a} = \frac{\sin B}{\sin b} = \frac{\sin C}{\sin c}$$

$$\cos a = \cos b \cos c + \sin b \sin c \cos A$$

$$\cos a = \frac{\cos b \cos(c \pm \theta)}{\cos(\theta)} \qquad \text{where } \tan \theta = \tan b \cos A$$

$$\cos A = -\cos B \cos C + \sin B \sin C \cos a$$

In solving spherical triangle problems we can use either the scientific keyboard function evaluation or, on the four-function calculators, the

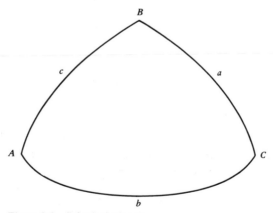

Figure 3-3 Spherical triangle.

polynomial or Chebyshev approximation to the transcendental functions involved. These developments for the right triangles are shown here not so much because of their unique form for pocket calculating, but because they are very frequently encountered in almost all forms of engineering analysis and, again, are provided here for the sake of completeness.

3-8 COMPLEX VARIABLES AND FUNCTIONS*

In the remainder of the book, and in Chapter 4 in particular, the equations and formulas used for analysis on the pocket calculator hold both for real and complex variables. In this section, we touch briefly on analysis with complex variables.

Complex variable analysis on the pocket calculator results in nothing more than keeping track of the real and imaginary coordinates either in polar or in Cartesian form. Pocket calculators with conversion from rectangular to polar make the analysis with complex variables particularly easy. Since virtually all advanced scientific calculators have this feature, we assume here that it is present. The formulas given here for analysis with complex variables and for the evaluation of functions of complex variables can be quickly and easily developed on the four-function calculator using the trigonometric functions developed earlier in this chapter or in Chapter 1.

The addition and subtraction of two complex variables are simply defined by

$$(x_1 + iy_1) + (x_2 + iy_2) = (x_1 + x_2) + i(y_1 + y_2)$$

$$(x_1 + iy_1) - (x_2 + iy_2) = (x_1 - x_2) + i(y_1 - y_2)$$

Multiplication is more conveniently done in polar coordinates; that is,

$$(x_1 + iy_1)(x_2 + iy_2) = r_1 r_2 e^{i(\theta_1 + \theta_2)}$$

where

$$x_1 + iy_1 = r_1 e^{i\theta_1}$$

$$x_2 + iy_2 = r_2 e^{i\theta_2}$$

Here r is the positive root sum square of the imaginary and real components of the complex number $v = (x^2 + y^2)^{1/2}$ and

$$\theta = \text{inverse tan} \frac{y}{x}$$

*See Appendix 3 and Appendix 4

The division of two complex numbers is given by

$$\frac{x_1 + iy_1}{x_2 + iy_2} = \frac{r_1}{r_2} e^{i(\theta_1 - \theta_2)}, \qquad (x_2 + iy_2 \neq 0)$$

Or, in rectangular form,

$$\frac{x_1 + iy_1}{x_2 + iy_2} = \frac{(x_1 + iy_1)(x_2 - iy_2)}{x_2^2 + y_2^2} = \frac{x_1 x_2 + y_1 y_2 + i(x_2 y_1 - x_1 y_2)}{x_2^2 + y_2^2}$$

where the denominator is developed by multiplying the numerator and denominator by the complex conjugate of the denominator.

We will frequently encounter certain commonly used functions of complex variables (complex functions). The most often occurring one is the modulus (absolute value) of a complex number, which is defined by

$$|x + iy| = (x^2 + y^2)^{1/2}$$

Another commonly encountered complex function is that of the square of a complex number, which is simply given by

$$(x + iy)^2 = (x^2 - y^2) + i(2xy)$$

Not so easily remembered but occasionally encountered is the square root of a complex number, which is given by

$$\sqrt{x + iy} = \pm \left[\left(\frac{x + (x^2 + y^2)^{1/2}}{2} \right)^{1/2} + \frac{iy}{2} \left(\frac{x + x(x^2 + y^2)^{1/2}}{2} \right)^{-1/2} \right]$$

Clearly, the powers and roots of a complex function can be more easily evaluated in polar coordinates. Thus, in general,

$$(x + iy)^n = r^n e^{in\theta} = r^n (\cos n\theta + i \sin n\theta)$$

and

$$(x + iy)^{1/n} = r^{1/n} \left[\cos\left(\frac{\theta + 2\pi k}{n} \right) + i \sin\left(\frac{\theta + 2\pi k}{n} \right) \right]$$

These formulas are written for angles in degrees, not radians, and for angles where n is an integer greater than zero and k takes on any integer values from zero $n-1$. The only restriction on these two complex functions is that the complex variable cannot equal zero.

Exponential and logarithmic functions of complex variables are also easily developed when the complex variable is written in polar form:

$$x + iy = re^{i\theta}$$

Then it is straightforward to develop

$$e^{(x+iy)} = e^x(\cos y + i \sin y)$$

Similarly, the natural log of a complex variable is given by

$$\ln(x + iy) = \ln r + i\theta \pm 2\pi ik, \qquad (k = 0, 1, 2, \cdots) \tag{3-4}$$

These relationships can be generalized to any base according to the relation

$$a^{(x+iy)} = e^{(x+iy)\ln a}$$

$$\log_a(x + iy) = \frac{\ln(x + iy)}{\ln a}$$

Even more generally, the complex powers, complex roots, and complex logarithms of a complex variable can be developed, again using polar coordinates. If we are given

$$z = x + iy \qquad \text{and} \qquad w = v + iv$$

then the complex powers and roots of a complex number are given by

$$z^w = e^{w \ln z} \tag{3-5}$$

$$z^{1/w} = e^{(\ln z)(1/w)} \tag{3-6}$$

Finally,

$$\log_z(w) = \frac{\ln(w)}{\ln(z)} \tag{3-7}$$

In equations 3-5, 3-6, and 3-7 we can use equation 3-4 for taking the natural log of a complex number.

Complex trigonometric functions are often presented in complex variable theory books more from the standpoint of derivation and development than from the standpoint of numerical evaluation. Hence the numerical evaluation formulas sometimes get buried in the derivation. Here we present the formulas for complex trigonometric functions in a form that is easily evaluated on the pocket calculator. No attempt is made to derive these formulas, interesting though they are, because again the emphasis here is on numerical evaluation of the functions.

The most straightforward expressions are the sine, cosine, and tangent functions of the complex variable

$$z = x + iy$$

Then

$$\sin z = \sin x \cosh y + i \cos x \sinh y$$

$$\cos z = \cos x \cosh y - u \sin x \sinh y$$

$$\tan z = \frac{\sin 2x + i \sinh 2y}{\cos 2x + \cosh 2y}$$

Less straightforward are the complex inverse trigonometric functions. Again, when

$$z = x + iy$$

the inverse sine of z is given by the relation

$$\sin^{-1} z = k\pi + (-1)^k \sin^{-1} \beta + (-1)^k i \operatorname{sgn}(y) \ln\left[\alpha + (\alpha^2 - 1)^{1/2}\right]$$

In this formula

$$\alpha = \tfrac{1}{2}\sqrt{(x+1)^2 y^2} + \tfrac{1}{2}\sqrt{(x-1)^2 + y^2} \qquad (3\text{-}8)$$

$$\beta = \tfrac{1}{2}\sqrt{(x+1)^2 + y^2} - \tfrac{1}{2}\sqrt{(x-1)^2 + y^2} \qquad (3\text{-}9)$$

and the function $\operatorname{sgn}(y)$ is given by the relation

$$\operatorname{sgn}(y) = \begin{cases} 1 & \text{if } y \geqslant 0 \\ -1 & \text{if } y < 0 \end{cases} \qquad (3\text{-}10)$$

Finally, k in this formula is an integer. A convenient simplification for pocket calculator analysis is to take into account the fact that the inverse trigonometric functions are multiple valued and thus the $k=0$ case (the easiest to evaluate numerically) can be used to evaluate the inverse sine of z when care is taken to account for the "quadrant" in which z is being determined. Then $\sin^{-1} z$ simplifies to the form

$$\sin^{-1} z = \sin^{-1} \beta + i \operatorname{sgn}(y) \ln\left[\alpha + (\alpha^2 - 1)^{1/2}\right]$$

Similarly the inverse cosine can be numerically evaluated from the equation

$$\cos^{-1}z = \cos^{-1}\beta - i\operatorname{sgn}(y)\ln\left[\alpha + (\alpha^2 - 1)^{1/2}\right] \qquad (3\text{-}11)$$

where α and β are given by equations 3-8 and 3-9 and $\operatorname{sgn}(y)$ is given by equation 3-10. Here, as before, k is assumed to be zero. However, were k not equal to zero the more general form of equation 3-11 is given by

$$\cos^{-1}z = 2k\pi \pm \left\{\cos^{-1}\beta - u\operatorname{sgn}(y)\ln\left[\alpha + (\alpha^2 - 1)^{1/2}\right]\right\}$$

Finally, the inverse tangent in its most general form is given by

$$\tan^{-1}z = \tfrac{1}{2}\left[(2k+1)\pi - \tan^{-1}\left(\frac{1+y}{x}\right) - \tan^{-1}\left(\frac{1-y}{x}\right)\right]$$
$$+ \frac{i}{4}\ln\left[\frac{(1+y)^2 + x^2}{(1-y)^2 + x^2}\right]$$

which when $k=0$ simplifies to the form

$$\tan^{-1}z = -\tfrac{1}{2}\left[\tan^{-1}\left(\frac{1+y}{x}\right) + \tan^{-1}\left(\frac{1-y}{x}\right)\right] + \frac{i}{4}\ln\left[\frac{(1+y)^2 + x^2}{(1-y)^2 + x^2}\right]$$

With these relationships it is a simple matter to define the complex hyperbolic and complex inverse hyperbolic functions in terms of the trigonometric functions and their inverses:

$$\sinh z = -i\sin iz$$

$$\cosh z = \cos iz$$

$$\tanh z = \frac{\sinh 2x + i\sin 2y}{\cosh 2x + \cos 2y}$$

Similarly, the inverse hyperbolic functions are defined as

$$\sinh^{-1}z = -i\sin^{-1}iz$$

$$\cosh^{-1}z = i\cos^{-1}z$$

$$\tanh^{-1}z = -i\tan^{-1}iz$$

Other complex trigonometric relationships useful in evaluation of complex functions are

$$\csc z = (\sin z)^{-1}$$

$$\sec z = (\cos z)^{-1}$$

$$\coton z = \frac{\sin 2x - i \sinh 2y}{\cosh 2y - \cos 2x}$$

$$\csc^{-1} z = \sin^{-1}(z^{-1})$$

$$\sec^{-1} z = \cos^{-1}(z^{-1})$$

$$\coton^{-1} z = \frac{\pi}{2} - \tan^{-1}(z)$$

$$\csch z = i \csc iz$$

$$\sech z = \sec iz$$

$$\coth z = \frac{\sinh 2x - i \sin 2y}{\cosh 2x - \cos 2y}$$

$$\csch^{-1} z = i\csc^{-1} iz$$

$$\sech^{-1} z = i\sec^{-1} z$$

$$\coth^{-1} z = i\cot^{-1} iz$$

3-9 REFERENCES

For this chapter refer to the *Handbook of Mathematical Functions*, U.S. Department of Commerce, National Bureau of Standards, Applied Mathematics Series 55, 1900.

CHAPTER 4

NUMERICAL EVALUATION
OF ADVANCED FUNCTIONS

4.1 INTRODUCTION

Even the simplest pocket calculator can evaluate *advanced mathematical functions* to accuracies required in engineering use and certainly carrying as many significant digits as do the typical tables in mathematical handbooks. In part, then, this chapter deals with freeing the analyst from having to carry or have access to extensive tables to numerically evaluate the advanced mathematical function. Among the advanced functions considered in this chapter are the exponential integral, the gamma function, the error function and Fresnal integrals, Legendre's polynomials, Bessel functions of integer and fractional orders, Confluent hypergeometric functions, Chebyshev polynomials, hypergeometric functions, Hermite polynomials, and Laguerre polynomials. Again, we stress not so much the analysis with these functions and their analytical properties as their numerical evaluation on the pocket calculator.

There are three methods for numerically evaluating advanced mathematical functions:

1. The function is approximated by a polynomial approximation or curve fit that permits accurate evaluation of the function directly through analytic substitution.

2. If the function is one of a sequence of *generated polynomials*, the low-order polynomials can be determined for the argument of the function and the higher-order polynomials then numerically evaluated by means of the recursion formulas.

3. Successive partial sums of the series that describes the advanced function are computed.

114

The first method has the greatest body of mathematical literature. It also is the simplest to apply in numerically evaluating advanced functions in that it involves the simple procedure of evaluating a polynomial. The third alternative is the least attractive, since rapidly converging series are often difficult to develop over all intervals of interest. An example of this is the Bessel function where a number of series can be written that converge quickly in certain intervals, but there is no one series that converges quickly over the entire range of the independent variable in the Bessel function. In fact, the Bessel function requires the same consideration in developing polynomial approximations. To get precision, polynomial approximation with a reasonable number of terms requires more than one polynomial approximation to span the interval of the independent variable from minus infinity to plus infinity.

Finally, the second approach (the use of low-order polynomials to determine the argument of the advanced function and then employing recursion formulas to numerically evaluate higher-order polynomials) is used extensively in the generation of accurate mathematical tables and thus is handy for pocket calculator analysis, though somewhat tedious at times.

In this chapter, all three approaches are used, each where appropriate for evaluating the advanced mathematical functions covered here. Care has been taken to select, from the number of available numerical methods for evaluating these functions, those methods that can be implemented with a minimum amount of work on pocket calculators, and particularly the four-function variety.

Where tradeoffs are difficult the method that leads to the quickest evaluation has been selected. A specific example would be the Chebyshev polynomials which are evaluated here using recursion formulas, rather than sine and cosine functions, which are available on most scientific calculators. The reason for this and a similar situation in the half-integer Bessel functions is that the approach presented here can be numerically calculated on the four-function calculator which does not have the sine and cosine functions.

Those who have done the evaluated certain mathematical functions on large-scale digital computers should recognize that the methods chosen here *are not* necessarily the same as those commonly used on large digital computers. Partly the numerical methods are chosen for the pocket calculator and partly because the methods are to be instructive to students and working engineers who may have been away from the application of these functions. For these cases, familiar numerical methods are often chosen, though they require slightly more work than the numerical methods used for large computer numerical evaluation. It is important to remember that in pocket calculator analysis a mathematical function is usually evaluated only a few times, while on a large digital computing machine it might be evaluated a great many times. For large machines, the emphasis is on maximizing the accuracy with the minimum number of steps involved in the subroutine. In pocket calculator analysis, the emphasis is more on understanding the method and providing accuracy consistent with the display in the pocket calculator, on a one-time basis. Thus the requirements for a numerical method for pocket calculator evaluation are significantly different than those for large computer evaluations.

4.2 EXPONENTIAL, SINE, AND COSINE INTEGRALS

Four commonly encountered integrals are two forms of the exponential integrals and the sine and cosine integrals. The exponential integrals that we discuss here are of the form

$$E_1(z) = \int_z^\infty \frac{e^{-t}}{t}\,dt, \qquad (|\arg z| < \pi)$$

$$Ei(x) = -\int_{-x}^\infty \frac{e^{-t}}{t}\,dt = \int_{-\infty}^x \frac{e^t}{t}\,dt, \qquad (x > 0)$$

More generally we are also interested in the exponential integral

$$E_n(z) = \int_1^\infty \frac{e^{-zt}}{t^n}\,dt, \qquad (n = 0, 1, 2, \ldots, \mathrm{Re}(z) > 0)$$

and in methods for numerically evaluating both

$$\alpha_n(z) = \int_1^\infty t^n e^{-zt}\,dt, \qquad (n = 0, 1, 2, \ldots, \mathrm{Re}(z) > 0)$$

$$\beta_n(z) = \int_{-1}^{+1} t^n e^{-zt}\,dt, \qquad (n = 0, 1, 2, \ldots)$$

Though each of these integrals is defined for complex arguments, our interest here is primarily with their evaluation for real arguments. However, it is usually assumed that the path of complex integration does not include the origin, nor does it cross the negative real axis.

The approach to numerically evaluating the exponential integrals is first to have a means to numerically evaluate them for $n = 1$, and then use recursion formulas for numerically evaluating the higher-order exponential integrals. There are two approaches to evaluating the integrals: the use of either an infinite series or rational polynomial approximations (see Chapter 8). The series expansions for these functions are

$$Ei(x) = \gamma + \ln x + \sum_{n=1}^{\infty} \frac{x^n}{nn!}, \qquad (x > 0)$$

$$E_1(z) = -\gamma - \ln z - \sum_{n=1}^{\infty} \frac{(-1)^n z^n}{nn!}, \qquad (|\arg z| < \pi)$$

$$E_n(z) = \frac{(-z)^{n-1}}{(n-1)!}[-\ln z + \psi(n)] - \sum_{\substack{m=0 \\ m \neq n-1}}^{\infty} \frac{(-z)^m}{(m-n+1)m!}, \qquad (|\arg z| < \pi)$$

$$\alpha_n(z) = n! z^{-(n+1)} e^{-z} \left(\sum_{i=0}^{n} \frac{z^i}{i!} \right)$$

$$\beta_n(z) = n! z^{-(n+1)} \left\{ e^z \left(\sum_{i=0}^{n} (-1)^i \frac{z^i}{i!} \right) - e^{-z} \left(\sum_{i=0}^{n} \frac{z^i}{i!} \right) \right\}$$

Here

$$\psi(1) = -\gamma$$

$$\psi(n) = -\gamma + \sum_{m=1}^{n-1} \frac{1}{m}$$

and

$$\gamma = 0.5772156649 \qquad \text{(Euler's constant)}$$

These functions can be evaluated more conveniently, using rational polynomial approximations.

For the dependent variable x on the interval zero to 1, the exponential integral can be evaluated with the polynomial

$$E_1(x) + \ln x = a_0 + x(a_1 + x(a_2 + x(a_3 + x(a_4 + a_5 x)))) + \epsilon(x)$$

with an accuracy of two parts in 10^7, using the coefficients

$$a_0 = -0.57721566 \qquad a_3 = 0.05519968$$
$$a_1 = 0.99999193 \qquad a_4 = -0.00976004$$
$$a_2 = -0.24991055 \qquad a_5 = 0.00107857$$

when x is on the interval

$$1 \leqslant x < \infty$$

Over the range where x is greater than or equal to 1 the rational polynomial approximation

$$xe^x E_1(x) = \frac{x^2 + a_1 x + a_2}{x^2 + b_1 x + b_2} + \epsilon(x)$$

using the coefficients

$$a_1 = 2.334733 \qquad b_1 = 3.330657$$
$$a_2 = 0.250621 \qquad b_2 = 1.681534$$

can evaluate the auxiliary exponential integral to an error of

$$|\epsilon(x)| < 5 \times 10^{-5}$$

For even greater precision, the exponential integral can be evaluated over the interval x greater than or equal to 10, using the same rational polynomial but with the coefficients

$$a_1 = 4.03640 \qquad b_1 = 5.03637$$
$$a_2 = 1.15198 \qquad b_2 = 4.19160$$

Here the error is given by the relation

$$|\epsilon(x)| < 10^{-7}$$

For x greater than 1, the rational polynomial approximation

$$xe^x E_1(x) = \frac{a_4 + x(a_3 + x(a_2 + x(a_1 + x)))}{b_4 + x(b_3 + x(b_3 + x(b_1 + x)))} + \epsilon(x)$$

using the coefficients

$$a_1 = 8.5733287401 \qquad b_1 = 9.573322454$$
$$a_2 = 18.0590169730 \qquad b_2 = 25.6329561486$$
$$a_3 = 8.6347608925 \qquad b_3 = 21.0996530827$$
$$a_4 = 0.2677737343 \qquad b_4 = 3.9584969228$$

can be used to evaluate the exponential integral to an accuracy of

$$|\epsilon(x)| < 2 \times 10^{-8}$$

Once the exponential integral is numerically evaluated, the following recursion formula can be used to compute the higher-order exponential integrals for the same arguments:

$$E_{n+1}(z) = \frac{1}{n}[e^{-z} - zE_n(z)], \qquad (n = 1, 2, 3, \ldots)$$

$$z\alpha_n(z) = e^{-z} + n\alpha_{n-1}(z), \qquad (n = 1, 2, 3, \ldots)$$

$$z\beta_n(z) = (-1)^n e^z - e^{-z} + n\beta_{n-1}(z), \qquad (n = 1, 2, 3, \ldots)$$

The sine and cosine integrals are defined as

$$Si(z) = \int_0^z \frac{\sin t}{t}\, dt$$

$$Ci(z) = \theta + \ln z + \int_0^z \frac{\cos(t) - 1}{t}\, dt, \qquad (|\arg z| < \pi)$$

Furthermore, we make note of the definition

$$si(z) = Si(z) - \frac{\pi}{2}$$

Then two auxiliary functions can be developed that have the form

$$f(z) = Ci(z)\sin z - si(z)\cos z$$
$$g(z) = -Ci(z)\cos z - si(z)\sin z$$

Then the sine and cosine integrals can be written in terms of the auxiliary function as

$$Si(z) = \frac{\pi}{2} - f(z)\cos z - g(z)\sin z$$

$$Ci(z) = f(z)\sin z - g(z)\cos z$$

where the auxiliary integrals are defined according to the relations

$$f(z) = \int_0^\infty \frac{\sin t}{t+z}\, dt$$

or

$$f(z) = \int_0^\infty \frac{e^{-zt}}{t^2+1}\, dt$$

and

$$g(z) = \int_0^\infty \frac{\cos t}{t+z}\, dt$$

or

$$g(z) = \int_0^\infty \frac{te^{-tz}}{t^2+1}\, dt$$

subject to the condition for convergence of these integrals

$$Re(z) > 0$$

The reason for doing this is that rational approximations to the auxiliary functions are easily developed with high precision.

For four-place precision, the auxiliary function can be determined using the rational approximation

$$f(x) = \frac{1}{x}\left(\frac{a_2 + x^2(a_1 + x^2)}{b_2 + x^2(b_1 + x^2)}\right) + \epsilon(x) \qquad g(x) = \frac{1}{x^2}\left(\frac{a_2 + x^2(a_1 + x^2)}{b_2 + x^2(b_1 + x^2)}\right) + \epsilon(x)$$

$$|\epsilon(x)| < 2 \times 10^{-4} \qquad\qquad\qquad\qquad |\epsilon(x)| < 10^{-4}$$

for x greater than 1 using the coefficients

$$a_1 = 7.241163 \qquad\qquad a_1 = 7.547478$$
$$a_2 = 2.463936 \qquad\qquad a_2 = 1.564072$$
$$b_1 = 9.068580 \qquad\qquad b_1 = 12.723684$$
$$b_2 = 7.157433 \qquad\qquad b_2 = 15.723606$$

For precision to five parts in 10^7, the auxiliary functions can be approximated with the rational approximations

$$f(x) = \frac{1}{x}\left(\frac{a_4 + x^2(a_3 + x^2(a_2 + x^2(a_1 + x^2)))}{b_4 + x^2(b_3 + x^2(b_2 + x^2(b_1 + x^2)))}\right)$$

$$g(x) = \frac{1}{x^2}\left(\frac{a_4 + x^2(a_3 + x^2(a_2 + x^2(a_1 + x^2)))}{b_4 + x^2(b_3 + x^2(b_2 + x^2(b_1 + x^2)))}\right)$$

for $1 \leqslant x$. The coefficients for $f(x)$ are given by

$$a_1 = 38.027264 \qquad b_1 = 40.021433$$
$$a_2 = 265.187033 \qquad b_2 = 322.624911$$
$$a_3 = 335.677320 \qquad b_3 = 570.236280$$
$$a_4 = 38.102495 \qquad b_4 = 157.105423$$

and for equation $g(x)$

$$a_1 = 42.242855 \qquad b_1 = 48.196927$$
$$a_2 = 302.757865 \qquad b_2 = 482.485984$$
$$a_3 = 352.018498 \qquad b_3 = 1114.978885$$
$$a_4 = 21.821899 \qquad b_4 = 449.690326$$

The infinite series for numerically evaluating these functions are *

$$Si(z) = \sum_{n=0}^{\infty} \frac{(-1)^n z^{2n+1}}{(2n+1)(2n+1)!}$$

$$Si(z) = \pi \sum_{n=0}^{\infty} J_{n+1/2}^2\left(\frac{z}{2}\right)$$

and for the cosine integral

$$Ci(z) = \gamma + \ln z + \sum_{n=1}^{\infty} \frac{(-1)^n z^{2n}}{2n(2n)!}$$

*We shall see that fractional Bessel functions can be conveniently evaluated on the scientific pocket calculator.

4.3 NUMERICAL EVALUATION OF THE GAMMA FUNCTION AND ITS RELATED FUNCTIONS

The gamma function is defined by Euler's integral

$$\Gamma(z) = \int_0^\infty t^{z-1} e^{-t}\, dt, \qquad (\text{Re}\, z > 0)$$

or

$$\Gamma(z) = k^z \int_0^\infty t^{z-1} e^{-kt}\, dt, \qquad (\text{Re}\, z > 0, \text{Re}\, k > 0)$$

Euler's formula for evaluating the gamma function is of the form

$$\Gamma(z) = \lim_{n \to \infty} \frac{n!\, n^z}{z(z+1)(z+2)\cdots(z+n)}, \qquad (z \neq 0, -1, -2\ldots)$$

He also gave an infinite product expression for evaluating the gamma function:

$$\frac{1}{\Gamma(z)} = z e^{\gamma z} \prod_{n=1}^{\infty} \left[\left(1 + \frac{z}{n}\right) e^{-z/n} \right]$$

where

$$\gamma = \lim_{m \to \infty} \left[1 + \frac{1}{2} + \frac{1}{3} + \frac{1}{4} + \cdots + \frac{1}{m} - \ln m \right] = 0.5772156649\cdots$$

This number is known as Euler's constant. Only a little analysis is involved to show that the gamma function is analytic and single valued over the entire complex plane except at the points $z = -n$ ($n = 0, 1, 2, \ldots,$) where its poles occur. The residues of these poles can be evaluated and are found to be

$$\frac{(-1)^n}{n!}$$

The reciprocal of the gamma function has zeros at the points $z = -n$ ($n = 0, 1, 2, \ldots,$). The recursion formula for computing the gamma function is given by the expression

$$\Gamma(z+1) = z\Gamma(z)$$

which is related to the factorial of z according to the relation

$$\Gamma(z+1) = z!$$

It follows that the gamma function propagates from gamma at $(1+z)$ to gamma at $(n+z)$ according to the relation

$$\Gamma(n+z)=(n-1+z)!=(n-1+z)(n-2+z)\cdots(1+z)z!$$

Another nice property of the gamma function, which is easily evaluated on the pocket calculator, is Gauss' multiplication formula

$$\Gamma(nz)=(2\pi)^{\left(\frac{1-n}{2}\right)}n^{nz-1/2}\prod_{k=0}^{n-1}\Gamma\left(z+\frac{k}{n}\right)$$

This formula contains the duplication and triplication formulas given as a part of the gamma function characteristics as special cases of this more general multiplication formula.

The gamma function, being related to the factorial of a number, is related to the binomial coefficient according to the relationship

$$\binom{z}{w}=\frac{z!}{w!(z-w)!}=\frac{\Gamma(z+1)}{\Gamma(w+1)\Gamma(z-w+1)}$$

It is apparent that the gamma function's relationship to the factorial makes it convenient to evaluating the gamma function on scientific pocket calculators that have the factorial key.

The gamma function can be evaluated in several ways. One is by a series expansion for expansion of $1/\Gamma$ according to the relationship

$$\frac{1}{\Gamma(z)}=\sum_{k=1}^{\infty}c_k z^k,\qquad (|z|<\infty)$$

where the coefficients to give an accuracy up to 10 places (the nominal register size we would expect in current and even some future pocket calculators) are tabulated in Table 4-1. The advantage to using this type of series expansion technique is that the interval over which the series converges is the entire real axis. Polynomial approximations can be used over more restricted intervals. Two such approximations are

$$\Gamma(x+1)=x!$$
$$=1+x\left(a_1+x\left(a_2+x\left(a_3+x\left(a_4+a_5x\right)\right)\right)\right)+\epsilon(x)$$

$$\Gamma(x+1)=x!$$
$$=1+x\left(b_1+x\left(b_2+x\left(b_3+x\left(b_4+x\left(b_5+x\left(b_6+x\left(b_7+b_8x\right)\right)\right)\right)\right)\right)\right)$$
$$+\epsilon(x)$$

Table 4-1 Coefficients in the Expansions $1/\Gamma(z) = \sum_{k=1}^{\infty} c_k z^k$

k	c_k	k	c_k
1	1.0000000000	11	0.0001280502
2	0.5772156649	12	−0.0000201348
3	−0.6558780715	13	−0.0000012504
4	−0.0420026350	14	0.0000011330
5	0.1665386113	15	−0.0000002056
6	−0.0421977345	16	0.0000000061
7	−0.0096219715	17	0.0000000050
8	0.0072189432	18	−0.0000000011
9	−0.0011651675	19	0.0000000001
10	−0.0002152416		

where the coefficients in the polynomials are

$$a_1 = -0.5748646 \qquad b_1 = -0.577191652$$

$$a_2 = 0.9512363 \qquad b_2 = 0.988205891$$

$$a_3 = -0.6998588 \qquad b_3 = -0.897056937$$

$$a_4 = 0.4245549 \qquad b_4 = 0.918206857$$

$$a_5 = -0.1010678 \qquad b_5 = -0.756704078$$

$$b_6 = 0.482199394$$

$$b_7 = -0.193527818$$

$$b_8 = 0.035868343$$

On both of these polynomial approximations the range of the variable x is greater than or equal to zero but less than or equal to 1. The former has an accuracy of five parts in 10^5 and the latter polynomial approximation is accurate to three parts in 10^7. Also, because of Stirling's formula for approximating $x!$, the gamma function can be related to Stirling's approximation according to the equation

$$\Gamma(az + b) \approx \sqrt{2\pi}\, e^{-az} (az)^{az+b-1/2} \qquad (|\arg z| < \pi, a > 0)$$

Again, Stirling's approximation is easy to evaluate on most scientific pocket calculators.

4-4 THE ERROR FUNCTION AND FRESNEL INTEGRALS

The error function and its complements are defined as

$$\text{erf } z = \frac{2}{\sqrt{\pi}} \int_0^z e^{-t^2} dt$$

and

$$\text{erfc } z = \frac{2}{\sqrt{\pi}} \int_z^\infty e^{-t^2} dt = 1 - \text{erf } z$$

The error function can be conveniently computed using the series expansion

$$\text{erf } z = \frac{2}{\sqrt{\pi}} e^{-z^2} \sum_{n=0}^\infty \frac{2^n}{1.3 \cdots (2n+1)} z^{2n+1}$$

In fact, it is common in computing the error function on large computers to compute successive partial sums of the series and terminate the evaluation when two consecutive partial sums are equal. The same approach can be taken on the pocket calculator, although the calculations are tedious. Here again we can use rational approximations to the error function such as

$$\text{erf } z = 1 - [t(a_1 + t(a_2 + a_3 t))]e^{-z^2} + \epsilon(z), \qquad (0 \leqslant z)$$

where

$$t = \frac{1}{1 + pz}$$

and the coefficients are

$$p = 0.47047$$

$$a_1 = 0.3480242$$

$$a_2 = -0.0958798$$

$$a_3 = 0.7478556$$

This approximation is good to about 2.5 parts in 10^{-5}. Accuracy to 1.5×10^{-5} error can be achieved with a slightly longer series, with the addition of two terms as

$$\text{erf } z = 1 - \left[t(a_1 + t(a_2 + t(a_3 + t(a_4 + a_5 t)))) \right] e^{-z^2} + \epsilon(z)$$

where

$$t = \frac{1}{1+pz}$$

and the coefficients are

$$p = 0.3275911$$

$$a_1 = 0.254829592$$

$$a_2 = -0.284496736$$

$$a_3 = 1.421413741$$

$$a_4 = -1.453152027$$

$$a_5 = 1.061405429$$

The Fresnel integrals are defined by the relationships

$$C(z) = \int_0^z \cos\left(\frac{\pi t^2}{2}\right) dt$$

$$S(z) = \int_0^z \sin\left(\frac{\pi t^2}{2}\right) dt$$

The Fresnel integrals can be computed using the series expansion

$$C(z) = \sum_{n=0}^{\infty} \frac{(-1)^n (\pi/2)^{2n}}{(2n)!(4n+1)} z^{4n+1}$$

$$S(z) = \sum_{n=0}^{\infty} \frac{(-1)^n (\pi/2)^{2n+1}}{(2n+1)!(4n+3)} z^{4n+3}$$

Fortunately, these series tend to converge quickly and can be evaluated effectively on the pocket calculator.

Finally, as might be expected, the Fresnel integrals can be computed in terms of sines and cosines directly, but modulated by the auxiliary functions $f(z)$ and $g(z)$:

$$C(z) = \frac{1}{2} + f(z)\sin\left(\frac{\pi z^2}{2}\right) - g(z)\cos\left(\frac{\pi z^2}{2}\right)$$

$$S(z) = \frac{1}{2} - f(z)\cos\left(\frac{\pi z^2}{2}\right) - g(z)\sin\left(\frac{\pi z^2}{2}\right)$$

The auxiliary functions are approximated to low accuracy according to

$$f(z) = \frac{1 + 0.926z}{2 + 1.792z + 3.104z^2} + \epsilon(z)$$

where

$$|\epsilon(z)| \leqslant 2 \times 10^{-3}$$

and

$$g(z) = \frac{1}{2 + 4.142z + 3.492z^2 + 6.670z^3} + \epsilon(z)$$

where

$$|\epsilon(z)| \leqslant 2 \times 10^{-3}$$

4-5 LEGENDRE FUNCTIONS

Legendre functions are defined in terms of the hypergeometric functions as

$$P_\nu^\mu(z) = \frac{1}{\Gamma(1-\mu)}\left[\frac{z+1}{z-1}\right]^{\mu/2} F\left(-\nu, \nu+1; 1-\mu; \frac{1-z}{2}\right)$$

where F is defined by the relationship

$$F(a,b;c;z) = \sum_{n=0}^{\infty} \frac{(a)_n (b)_n}{(c)_n}\left(\frac{z^n}{n!}\right)$$

$$= \frac{\Gamma(c)}{\Gamma(a)\Gamma(b)} \sum_{n=0}^{\infty} \frac{\Gamma(a+n)\Gamma(b+n)}{\Gamma(c+n)}\left(\frac{z^n}{n!}\right)$$

This Legendre function is one of two that satisfy the differential equation

$$(1-z^2)\frac{d^2w}{dz^2} - 2z\frac{dw}{dz} + \left[\nu(\nu+1) - \frac{\mu^2}{1-z^2}\right]w = 0$$

The Legendre function of the second kind is defined by the equation

$$Q_\nu^\mu(z) = e^{i\mu\pi}2^{-(\nu+1)}\pi^{1/2}\frac{\Gamma(\nu+\mu+1)}{\Gamma(\nu+3/2)}z^{-(\nu+\mu+1)}(z^2-1)^{\mu/2}$$

$$\times F\left(1+\frac{\nu}{2}+\frac{\mu}{2}, \frac{1+\nu+\mu}{2}; \nu+\frac{3}{2}; \frac{1}{z^2}\right) \qquad (|z|>1)$$

where F is again the hypergeometric function. These formidable-looking expressions are easily evaluated on the pocket calculator using the recurrence relationships of varying order and degree:

$$(\nu-\mu+1)P_{\nu+1}^\mu(z) = (2\nu+1)zP_\nu^\mu(z) - (\nu+\mu)P_{\nu-1}^\mu(z)$$

$$P_\nu^{\mu+1}(z) = (z^2-1)^{-1/2}\{(\nu-\mu)zP_\nu^\mu(z) - (\nu+\mu)P_{\nu-1}^n(z)\}$$

$$P_{\nu+1}^\mu(z) = P_{\nu-1}^\mu(z) + (2\nu+1)(z^2-1)^{1/2}P_\nu^{\mu-1}(z)$$

The Legendre functions of both the first and second kind satisfy these same recurrence relations. The starting values for these recurrence relations are $P_0(x)=1$ and $P_1(x)=x$ and

$$Q_0(z) = \tfrac{1}{2}\ln\left(\frac{z+1}{z-1}\right), \qquad\qquad Q_0(x) = \tfrac{1}{2}\ln\left(\frac{1+x}{1-x}\right),$$

$$Q_1(z) = \frac{z}{2}\ln\left(\frac{z+1}{z-1}\right) - 1, \qquad Q_1(x) = \frac{x}{2}\ln\left(\frac{1+x}{1-x}\right) - 1$$

Here n is nonnegative and an integer.

Another approach to evaluating Legendre functions of integral order is to use Rodrigues' formula to generate the Legendre polynomials and then to write them in nested parenthetical form and evaluate them like any polynomial. Though plausible, this approach is not developed here because the numerical evaluation of the Legendre polynomials is quickly done using the recurrence formulas. For the sake of completeness, however, the other approach can be developed using the relationship

$$P_n(z) = \frac{1}{2^n n!} \frac{d^n (z^2 - 1)^n}{dz^n}$$

and

$$Q_n(x) = \frac{P_n(x)}{2} \ln\left(\frac{1+x}{1-x}\right) - W_{n-1}(x)$$

where

$$W_{n-1}(x) = \frac{2n-1}{n} P_{n-1}(x) + \frac{2n-5}{3(n-1)} P_{n-3}(x) + \frac{2n-9}{5(n-2)} P_{n-5}(x) + \cdots$$

$$W_{n-1}(x) = \sum_{m=1}^{n} \frac{1}{m} P_{m-1}(x) P_{n-m}(x)$$

and

$$W_{-1}(x) = 0$$

The derivative of Legendre polynomials of the first kind can be numerically evaluated using the recurrence relations

$$P_n'(x) = \frac{n+1}{1-x^2} [x P_n(x) - P_{n+1}(x)]$$

4-6 BESSEL FUNCTIONS

Bessel functions are solutions to the differential equation

$$z^2 \frac{d^2 w}{dz^2} + z \frac{dw}{dz} + (z^2 - \nu^2) w = 0$$

Of the three kinds of Bessel functions, the first is:

$$\mathbf{J} \pm \nu(z) = \left(\frac{z}{2}\right)^\nu \sum_{k=0}^{\infty} \frac{\left(-\frac{z^2}{4}\right)^k}{k! \Gamma(\nu + k + 1)}$$

The second is written as

$$\mathbf{Y}_\nu(z)$$

and the third as

$$H_{\nu}^{(1)}, \qquad H_{\nu}^{(2)}(z)$$

Bessel functions of the first type can also be expressed as hypergeometric functions and as an integral:

$$J_{\nu}(z) = \frac{(z/2)^{\nu}}{\Gamma(\nu+1)} {}_0F_1\left(\nu+1; -\frac{3^2}{4}\right)$$

$$J_{\nu}(z) = \left(\frac{1}{2}z\right)^{\nu}\left[\pi^{1/2}\Gamma\left(\nu+\frac{1}{2}\right)\right]^{-1}\int_0^{\pi}\cos(z\cos\theta)\sin^{2\nu}\theta\,d\theta$$

Here ${}_0F_1$ is the generalized hypergeometric function. Bessel's functions of the second and third types are written in terms of Bessel function of the first type to simplify their numerical evaluation according to the relationships

$$Y_{\nu}(z) = \frac{J_{\nu}(z)\cos(\nu\pi) - J_{-\nu}(z)}{\sin(\nu\pi)}$$

$$H_{\nu}^{(1)}(z) = J_{\nu}(z) + iY_{\nu}(z)$$

$$H_{\nu}^{(2)}(z) = J_{\nu}(z) - iY_{\nu}(z)$$

The numerical evaluation of these Bessel functions is somewhat involved if done analytically. All can be computed, however, using the recurrence relation

$$J_{\nu+1} = \frac{2\nu}{z}J_{\nu} - J_{\nu-1}$$

where the numerical values for the Bessel functions that go into this recurrence formula are given by polynomial approximations. Unfortunately, the startup behavior of the Bessel function for x from -3 to $+3$ is significantly different from the Bessel function evaluated for x greater than 3. Two levels of polynomial approximation are therefore involved.

The polynomial approximations for the Bessel functions of the first and second kinds that can be used in combination with recurrence formulas for ν equal to zero and 1 are shown in Tables 4-2 and 4-3.

Table 4-2 Polynomial Approximation of $J_0(x)$

● *On the interval* $-3 \leqslant x \leqslant +3$

$$J_0(x) = 1 - 2.2499997\left(\frac{x}{3}\right)^2 + 1.2656208\left(\frac{x}{3}\right)^4 - 0.3163866\left(\frac{x}{3}\right)^6$$

$$+ 0.0444479\left(\frac{x}{3}\right)^8 - 0.0039444\left(\frac{x}{3}\right)^{10} + 0.0002100\left(\frac{x}{3}\right)^{12} + \epsilon$$

Here $|\epsilon| < 5 \times 10^{-8}$

● *On the interval* $3 \leqslant x$

$$J_0(x) = x^{-1/2} f_0 \cos \theta_0$$

where

$$f_0 = 0.79788456 - 0.00000077(3/x) - 0.00552740(3/x)^2 - 0.00009512(3/x)^3$$

$$+ 0.00137237(3/x)^4 - 0.00072805(3/x)^5 + 0.00014476(3/x)^6 + \epsilon$$

Here $|\epsilon| < 1.6 \times 10^{-8}$

and

$$\theta_0 = x - 0.78539816 - 0.04166397(3/x) - 0.00003954(3/x)^2 + 0.00262573(3/x)^3$$

$$- 0.00054125(3/x)^4 - 0.00029333(3/x)^5 + 0.00013558(3/x)^6 + \epsilon$$

Here $|\epsilon| < 7 \times 10^{-8}$

Spherical Bessel functions, often called Bessel functions of fractional order, satisfy the modified Bessel differential equation

$$z^2 w'' + 2zw' + [z^2 - n(n+1)]w = 0, \qquad (n = 0, \pm 1, \pm 2, \cdots)$$

Spherical Bessel functions of the first kind are of the form

$$j_n(z) = \sqrt{\frac{\pi}{2z}} \, J_{n+1/2}(z)$$

Table 4-3 Polynomial Approximation of $J_1(x)$

● *In the interval* $-3 \leqslant x \leqslant +3$

$$x^{-1}J_1(x) = \tfrac{1}{2} - 0.56249985(x/3)^2 + 0.21093573(x/3)^4 - 0.03954289(x/3)^6$$

$$+ 0.00443319(x/3)^8 - 0.00031761(x/3)^{10} + 0.00001109(x/3)^{12} + \epsilon$$

Here $|\epsilon| < 1.3 \times 10^{-8}$

● *In the interval* $3 \leqslant x$

$$J_1(x) = x^{1/2} f_1 \cos \theta_1$$

where

$$f_1 = 0.79788456 + 0.00000156(3/x) + 0.01659667(3/x)^2 + 0.00017105(3/x)^3$$

$$- 0.00249511(3/x)^4 + 0.00113653(3/x)^5 - 0.00020033(3/x)^6 + \epsilon$$

Here $|\epsilon| < 4 \times 10^{-8}$

and

$$\theta_1 = x - 2.35619449 + 0.12499612(3/x) + 0.00005650(3/x)^2 - 0.00637879(3/x)^3$$

$$+ 0.00074348(3/x)^4 + 0.00079824(3/x)^5 - 0.00029166(3/x)^6 + \epsilon$$

Here $|\epsilon| < 9 \times 10^{-8}$

---

and those of the second kind take the form

$$y_n(z) = \sqrt{\frac{\pi}{2z}}\ Y_{n+1/2}(z)$$

The spherical Bessel functions of the third kind are given by

$$h_n^{(1)}(z) = j_n(z) + iy_n(z) = \sqrt{\frac{\pi}{2z}}\ H_{n+1/2}^{(1)}(z)$$

$$h_n^{(2)}(z) = j_n(z) - iy_n(z) = \sqrt{\frac{\pi}{2z}}\ H_{n+1/2}^{(2)}(z)$$

They can be numerically evaluated using the series

$$j_n(z) = \frac{z^n}{1.3.5\cdots(2n+1)}\left\{1 - \frac{z^2/2}{1!(2n+3)} + \frac{(3^2/2)^2}{2!(2n+3)(2n+5)} - \cdots\right\}$$

$$y_n(z) = -\frac{1.3.5\cdots(2n-1)}{z^{n+1}}\left\{1 - \frac{3^2/2}{1!(1-2n)} + \frac{(3^2/2)^2}{2!(1-2n)(3-2n)} - \cdots\right\}$$

where $n = 0, 1, 2, \cdots$ .Or they can be written out and numerically evaluated from the expansion of the spherical Bessel functions, using Rayleigh's generating formulas

$$j_n(z) = z^n\left(-\frac{1}{z}\frac{d}{dz}\right)^n\left(\frac{\sin z}{z}\right)$$

$$y_n(z) = -z^n\left(-\frac{1}{z}\frac{d}{dz}\right)^n\left(\frac{\cos z}{z}\right)$$

A simplification for numerically evaluating spherical Bessel functions of high order is to evaluate them at low order, say zero and 1, and compute the high-order spherical Bessel functions for the same argument, using the recurrence relations

$$j_{n+1} = \frac{(2n+1)}{z}j_n - j_{n-1}$$

This recursion formula applies to all four spherical Bessel functions.

An even simpler approach for calculators with sine and cosine functions using recurrence formulas is to evaluate $j_n(z)$ as

$$j_n(z) = f_n(z)\sin z + (-1)^{n+1}f_{-(n+1)}(z)\cos z$$

where f_n is generated with

$$f_{n+1} = \frac{(2n+1)}{z}f_n - f_{n-1}$$

using the starting values

$$f_0 = z^{-1} \quad \text{and} \quad f_1 = z^{-2}$$

4-7 THE CONFLUENT HYPERGEOMETRIC FUNCTION

The confluent hypergeometric function, usually written in the form

$$M(a,b,z) = 1 + \frac{az}{b} + \frac{(a)_2 z^2}{(b)_2 z!} + \cdots + \frac{(a)_n z^n}{(b)_n n!}$$

where

$$(b)_n = b(b+1)(b+2)(b+3) \cdots (b+n-1)$$

$$(a)_n = a(a+1)(a+2)(a+3) \cdots (a+n-1)$$

and

$$(a)_0 = 1 = (b)_0$$

is the solution to Kummers' differential equation

$$z \frac{d^2 w}{dz^2} + (b-z) \frac{dw}{dz} - aw = 0$$

The confluent hypergeometric function is evaluated on the pocket calculator directly as written by computing successive partial sums of this series. Similarly, the hypergeometric functions defined by the relation

$$F(a,b;c;z) = \sum_{n=0}^{\infty} \frac{(a)_n (b)_n}{(c)_n} \left(\frac{z^n}{n!} \right)$$

must be evaluated directly by computing successive partial sums of the series. In general, when the terms in the series do not change, enough terms have been taken for the numerical evaluation to be complete. It is apparent that the series is not defined if $c = -m(m = 0, 1, 2, \cdots)$, except when

$$b = -n(n = 0, 1, 2, \cdots)$$

where n must be less than m. It is worth mentioning here that the hypergeometric function can be used to initialize certain recursion formulas for other advanced functions in that the hypergeometric function is related to many of the orthogonal polynomials.

4-8 CHEBYSHEV, HERMITE, AND LAGUERRE POLYNOMIALS

Chebyshev polynomials can be easily evaluated numerically using the recurrence equations

$$T_{n+1}(x) = 2x T_n(x) - T_{n-1}(x)$$

where the starting values for the Chebyshev polynomials are $T_0(x) = 1$ and $T_1(x) = x$.

Hermite polynomials and Laguerre polynomials can also be evaluated using recurrence equations and initial values of the polynomials. For example, the Hermite polynomial can be numerically evaluated using the recurrence formula

$$H_{n+1}(x) = 2xH_n(x) - 2nH_{n-1}(x)$$

where the starting values are computed from the first two Hermite polynomials $H_0 = 1$ and $H_1 = 2x$. Again, n is a nonnegative integer.

The numerical evaluation of the Laguerre polynomial is found using the recurrence equation

$$L_{n+1}(x) = \frac{[(2n+1-x)L_n(x) - nL_{n-1}(x)]}{n+1}$$

Here the starting Laguerre polynomials are $L_0 = 1$ and $L_1 = 1 - x$. As before, n is a nonnegative integer.

It is worth repeating that to numerically evaluate advanced mathematical functions one of three approaches is usually employed:

1. The function is approximated by a polynomial approximation or curve fits that permit accurate evaluation of the function directly through analytic substitution.

2. If the function is one of a sequence of generated polynomials, the low-order polynomials can be determined for the argument of the function, and then recursion formulas used to numerically evaluate the higher-order polynomials.

3. The third alternative is simply to compute successive partial sums of the series that describes the advanced function.

Of the three, (1) is the most straightforward approach to evaluating the advanced functions. The least attractive approach is that presented in (3)—direct evaluation of the series approximation to the function. And, finally, a reasonable tradeoff between analytic substitution as described in (1) and direct series evaluation presented in (3) is the use of recurrence formulas to numerically evaluate high-order functions where the function is one of a set of sequences of functions developed with a generating formula.

4-9 REFERENCES

For this chapter consult the *Handbook of Mathematical Functions*, U.S. Department of Commerce, National Bureau of Standards, Applied Mathematics Series 55, 1900.

PART THREE

ADVANCED ANALYSIS
ON THE POCKET CALCULATOR

CHAPTER 5

FOURIER ANALYSIS

5-1 INTRODUCTION

We now turn to the Fourier analysis of discrete and continuous functions. Unlike in Chapter 4, where we talked about the numerical evaluation of advanced mathematical functions rather than their interpretation, here we also try to understand the results that can be obtained from pocket calculator evaluation of the discrete Fourier transform. Such issues as the relationships between the discrete spectrum of discrete functions and the discrete spectrum of continuous functions are discussed. The aliasing concept is examined to aid those not familiar with it in understanding the spectrum of sampled-data functions. In a very real sense, the pocket calculator can be a valuable teaching aid in frequency-domain analysis in that it permits the quick evaluation of the spectra associated with discrete functions which (when sampled at sufficiently high frequency) approximate continuous functions. For this reason, and to permit a quick evaluation of spectra in practical analysis, the formulas for 12-point and 24-point discrete Fourier spectra are given and the procedures for their quick evaluation on the pocket calculator are presented. Those using a simple four-function calculator will find the 12-point formula of particular interest, since no evaluation of sines or cosines is required for a determination of the discrete spectrum of a sequence of sampled values.

Finally, we discuss function reconstruction, using pocket calculators with scientific keyboards, and procedures for Fourier series evaluation.

5-2 THE FOURIER SERIES OF CONTINUOUS FUNCTIONS

The Fourier series of a continuous periodic function whose period is L is given by

$$F(x) = \frac{a_0}{2} + \left[a_1 \cos\left(\frac{2\pi x}{L} \right) + a_2 \cos\left(\frac{4\pi x}{L} \right) + a_3 \cos\left(\frac{6\pi x}{L} \right) + \cdots \right]$$

$$+ \left[b_1 \sin\left(\frac{2\pi x}{L} \right) + b_2 \sin\left(\frac{4\pi x}{L} \right) + b_3 \sin\left(\frac{6\pi x}{L} \right) + \cdots \right] \qquad (5\text{-}1)$$

where the coefficients a_k and b_k are given by

$$a_k = \frac{2}{L} \int_0^L F(x) \cos\left(\frac{2k\pi x}{L} \right) dx, \qquad (k = 0, 1, 2, \cdots) \qquad (5\text{-}2)$$

$$b_k = \frac{2}{L} \int_0^L F(x) \sin\left(\frac{2k\pi x}{L} \right) dx, \qquad (k = 1, 2, \cdots) \qquad (5\text{-}3)$$

This series has an infinite number of terms. There are an infinite number of coefficients defined by equations 5-2 and 5-3. Physically, the fundamental and an infinity of its harmonics can determine an infinite sequence of coefficients for the sines and cosines in the series expansion of equation 5-1.

There are problems with the convergence of Fourier series, which are usually associated with special functions not frequently encountered in engineering analysis. Since the sines and cosines that make up the Fourier series are orthogonal functions, the coefficients of its components can be written in polar form. In this case the modulus of the radius vector can be plotted as a function of frequency, as well as the phase angle of the resultant. These form the amplitude and phase shift curves used by control systems analysts to examine the stability of feedback control processes.

Finally, the power spectrum of a function approximated with a Fourier series can be prepared by plotting the square of the modulus of the Fourier coefficients in polar form. Since this is equivalent to summing the squares of the coefficients for the sine and cosine functions at the same frequency, it is apparent that the phase information is lost when the power spectrum is presented.

Inspection of the Fourier series reveals that if the function $F(x)$ is moved vertically along the ordinate, only the value of a_0 is changed. This can readily be seen by noting in equation 5-2 that when $k=0$ the coefficient becomes the average value of $F(x)$ on the interval zero to L. Note that if we were to set $F_1(x) = F(x) + a_0/2$ in equation 5-1, the Fourier series of $F_1(x)$ would have no average component (DC component). It is also apparent that if $F(x)$ is translated horizontally along the abscissa, the distribution of weight among the coefficients in the Fourier series changes. In particular, for any given frequency component, the distribution among the sine and cosine terms is different. However, the sum of their squares remains invariant. From this we see that the power spectrum of a Fourier series is invariant with respect to translation of the function $f(x)$ along the abscissa (assuming that $f(x)$ is periodic).

What should be kept in mind about the Fourier series approximation of a function is that the function being approximated with the Fourier series is assumed to be *periodic.* Though it can have discontinuities over the period, the function must be periodic. For the student, what might seem counterintuitive at first is that this function defined over a finite interval in the domain of the real numbers has an infinite number of lines in its discrete spectrum defined over the infinite domain of the discrete frequencies. The reason for this lies in the nature of the Fourier series. The fundamental frequency component for the finite length record is set by the length of the record itself. Then the harmonics that make up the rest of the series are determined by pairs of sines and cosines that fit with multiple oscillations over the period of the periodic function, and there are an infinity of these sines and cosines.

5-3 THE FOURIER SERIES OF DISCRETE FUNCTIONS

Now let us consider a discrete function defined by a set of equally spaced discrete values of the function Y defined at equally spaced values of the domain of the independent variable. Following Hamming, we consider only an even number of points, $2N$. In what follows, the $2N$ sample points are

$$0, \frac{L}{2N}, \frac{2L}{2N}, \frac{3L}{2N}, \dots, \frac{(2N-1)L}{2N} \tag{5-4}$$

which can be written as

$$x_n = \frac{nL}{2N}, \qquad (n=0, 1, \dots, 2N-1) \tag{5-5}$$

The Fourier series expansion of an arbitrary function $F(x)$ defined on the set of points x_n can be written as

$$F(x) = \frac{A_0}{2} + \sum_{k=1}^{N-1}\left[A_k\cos\left(\frac{2k\pi x}{L}\right) + B_k\sin\left(\frac{2k\pi x}{L}\right)\right] + \frac{A_N}{2}\cos\left(\frac{2N\pi x}{L}\right)$$

$$\tag{5-6}$$

where

$$A_k = \frac{1}{N} \sum_{n=0}^{2N-1} F(x_n)\cos\left(\frac{2k\pi x_n}{L}\right), \qquad (k=0,1,\ldots,N) \qquad (5\text{-}7)$$

$$B_k = \frac{1}{N} \sum_{n=0}^{2N-1} F(x_n)\sin\left(\frac{2k\pi x_n}{L}\right), \qquad (k=0,1,\ldots,N-1) \qquad (5\text{-}8)$$

An interesting aspect of Fourier series is that the sine and cosine functions are not only defined and orthogonal over a continuous interval of the dependent variable but are also orthogonal on any set of equally spaced discrete points on the same interval. This is important for the numerical evaluation of the frequency components of the discrete functions in that we are usually only given samples of the function on the set of equally spaced points. One might expect that the coefficients for the continuous Fourier series could then be approximated using numerical integration. Although this, in fact, can be done, equations 5-7 and 5-8 show that the coefficients for the Fourier spectrum of a discrete function can be computed exactly without involving numerical integration approximation.

It is apparent from equations 5-7 and 5-8 that this expansion will have only $2N$ terms, rather than an infinity of terms, as is characteristic of the Fourier series approximation of continuous functions. Also, the frequency spectrum for the discrete Fourier transform will have only half as many lines as sampled values. Thus, if there are 10 sampled values, the power spectrum will have only five discrete lines. Clearly, a question that needs to be answered is, 'what happened to the rest of the infinity of components of the spectrum of the continuous function $F(x)$? Another relevant and equally important question is, how can the discrete spectrum be constructed if we know the continuous spectrum and the number of sampled values? Conversely, given the discrete spectrum of the discrete function $f(n\Delta x)$, what can we tell about the discrete spectrum of the continuous function $f(x)$?

5-4 THE RELATIONS BETWEEN THE FOURIER SERIES EXPANSION OF DISCRETE AND CONTINUOUS FUNCTIONS

Only a little analysis will show that the spectral components of a discrete function are related to the spectral components of its continuous function counterpart according to the relationship

$$A_k = a_k + \sum_{m=1}^{\infty} (a_{2Nm-k} + a_{2Nm+k}) \qquad (5\text{-}9)$$

$$B_k = b_k + \sum_{m=1}^{\infty} (b_{2Nm+k} - b_{2Nm-k}) \qquad (5\text{-}10)$$

The constant term A_0 is given by

$$A_0 = a_0 + 2 \sum_{m=1}^{\infty} a_{2Nm} \qquad (5\text{-}11)$$

Let us examine equation 5-9. The first six terms of an even function discrete spectrum (say, for $2N = 10$ seconds) when written out take the following form:

Spectrum Component	Discrete Function Spectrum from Continuous Function Spectrum
$2\alpha_{\omega=0} =$	$A_0 = a_0 + 2(a_{10} + a_{20} + a_{30} + a_{40} + \cdots)$
$\alpha_{\omega=1\,Hz} =$	$A_1 = a_1 + (a_9 + a_{11}) + (a_{19} + a_{21}) + (a_{29} + a_{31}) + \cdots$
$\alpha_{\omega=2\,Hz} =$	$A_2 = a_2 + (a_8 + a_{12}) + (a_{18} + a_{22}) + (a_{28} + _{32}) + \cdots$
$\alpha_{\omega=3\,Hz} =$	$A_3 = a_3 + (a_7 + a_{13}) + (a_{17} + a_{23}) + (a_{27} + a_{33}) + \cdots$
$\alpha_{\omega=4\,Hz} =$	$A_4 = a_4 + (a_6 + a_{14}) + (a_{16} + a_{24}) + (a_{26} + a_{34}) + \cdots$
$2\alpha_{\omega=5\,Hz} =$	$A_5 = a_5 + (a_5 + a_{15}) + (a_{15} + a_{25}) + (a_{25} + a_{35}) + \cdots$

Note: $B_k = 0 = b_k$ for even functions.

The factors of 2 at end points of the spectrum are due to the form of equation 5-6. Each of the discrete system coefficients when written out includes the effect of an infinite number of terms associated with the continuous spectrum. This somewhat stunning finding shows that any sizable power in the high-frequency components of the continuous-function spectrum will have the effect of those components appearing at low frequencies in the discrete-function spectrum. In the example chosen where the $2N$ is equal to 10 seconds, the lowest-frequency component in the discrete spectrum (besides the zero-frequency component) includes the amplitude of the 9 and 11 Hz frequency component in the Fourier series expansion of $F(x)$, as well as the 19th and 21st, the 29th and 31st, 39th and 41st, 49th and 51st, and so on.

Another *physical* way to think of this is that, had we been sampling an 11 Hz sine wave at 10 samples per second, the discrete function would have had spectral components at 1 Hz whose amplitude would be the *same* as that of the 11 Hz sine wave in the continuous spectrum. In general the frequencies present in the original continuous function $F(x)$ are summed together as a result of the sampling operation. In other words, the mere operation of sampling a continuous function (i.e., a discrete function defined only at integer multiples of the sampling period) results in the *folding* of the frequency spectrum around the information frequency (sampling frequency ω_s divided by 2), resulting in the folding of the high-frequency components of the continuous-function spectrum into the low-frequency components of the discrete-function spectrum. This effect is called "aliasing" (i.e., 1 Hz is the alias of 9 Hz when sampled at 10 Hz). Note that once the sampling process has taken place, its effect on the continuous-function spectrum cannot be undone. In our example, only when $a_k = 0$ for $k > 5$ will $A_k = a_k$. If $a_k \neq 0$ for $k > 5$, a_k cannot be determined by examining A_k! Finally, note that the highest-frequency component is $\omega_s/2$.

These formulas provide the means for computing the discrete-function spectrum directly from the continuous-function spectrum if it is known. The *CRC Standard Mathematical Tables* has a sizable table of Fourier series representations of commonly encountered functions and can be used to compute the discrete-function spectrum according to equations 5-9, 5-10, and 5-11 using a pocket calculator.

5-5 THE NUMERICAL EVALUATION OF THE FOURIER COEFFICIENTS

The coefficients of the discrete-function Fourier series expansion, previously given by equations 5-7, and 5-8, are recopied here for convenience.

$$A_k = \frac{1}{N} \sum_{n=0}^{2N-1} F(x_n)\cos\left(\frac{2k\pi x_n}{L}\right), \qquad (k=0,1,\ldots,N) \qquad (5\text{-}7)$$

$$B_k = \frac{1}{N} \sum_{n=0}^{2N-1} F(x_n)\sin\left(\frac{2k\pi x_n}{L}\right), \qquad (k=0,1,\ldots,N-1) \qquad (5\text{-}8)$$

These coefficients may be numerically evaluated, using recursion formulas. The procedure is as follows:

Step 1 Prepare a table of values U_m using the recursion relations

$$U_m = \left(2\cos\frac{\pi k}{N}\right)U_{m-1} - U_{m-2} + F(2N-m) \qquad \text{for} \quad m=2,3,4,\ldots,2N-1$$

(5-17)

where $U_0 = 0$ and $U_1 = F(2N-1)$

Step 2 Evaluate the coefficients of the cosine terms in the series, using the equation

$$A_k = \frac{1}{N}\left\{\left(\cos\frac{\pi k}{N}\right)U_{2N-1} - U_{2N-2} + F(0)\right\}$$

(5-18)

Step 3 Compute the coefficients for the sine terms in the series expansion, using the recursion formula

$$B_k = \frac{1}{N}\left(\sin\frac{\pi k}{n}\right)U_{2N-1}$$

(5-19)

While evaluation according to these formulas takes longer than does the Cooley-Tukey fast Fourier transform algorithm, for the low-order Fourier series analysis that can be conveniently done on the pocket calculator this recursion formula method involves only slightly fewer operations than the Cooley-Tukey algorithm. Moreover, the use of recursion formulas in the numerical evaluation of functions is efficiently done on the pocket calculator.

Hamming presents a convenient 12-point formula for Fourier analysis in his book *Numerical Methods for Scientists and Engineers*. First, the table of discrete values of the function F is written in an array ($A1$):

$$\left.\begin{array}{ccccccc} F(0) & F(1) & F(2) & F(3) & F(4) & F(5) & F(6) \\ & F(11) & F(10) & F(9) & F(8) & F(7) \end{array}\right\} A1$$

From this array we can compute a sequence of S's and T's by adding and subtracting, respectively, the two lines in the array $A1$ to form an array ($A2$) of S's and T's:

$$\left.\begin{array}{cccccccc} \text{Sums}\rightarrow & S(0) & S(1) & S(2) & S(3) & S(4) & S(5) & S(6) \\ \text{Differences}\rightarrow & & T(1) & T(2) & T(3) & T(4) & T(5) \end{array}\right\} A2$$

Then, rewriting array ($A2$) as

$$
\left.
\begin{array}{ccccccc}
S(0) & S(1) & S(2) & S(3) & T(1) & T(2) & T(3) \\
S(6) & S(5) & S(4) & & & T(5) & T(4)
\end{array}
\right\} A3
$$

we form array ($A4$) of U's, V's, P's, and Q's by adding and subtracting, respectively, the second line from the first line in array $A3$:

$$
\left.
\begin{array}{clccccccc}
\text{Sums}\rightarrow & U(0) & U(1) & U(2) & U(3) & P(1) & P(2) & P(3) \\
\text{Differences}\rightarrow & V(0) & V(1) & V(2) & & & Q(1) & Q(2)
\end{array}
\right\} A4
$$

The coefficients associated with the six discrete frequencies that make up the Fourier series representation of the 12-point discrete function can now be developed. First compute

$$
\begin{aligned}
\alpha_1 &= V(0) + \frac{V(2)}{2} & \beta_1 &= \frac{P(1)}{2} + P(3) \\
\alpha_2 &= U(0) + U(3) & \beta_2 &= P(1) - P(3) \\
\alpha_3 &= U(1) + U(2) & \beta_3 &= \frac{\sqrt{3}}{2}(Q(1) + Q(2)) \\
\alpha_4 &= \frac{\sqrt{3}}{2} V(1) & \beta_4 &= \frac{\sqrt{3}}{2}(Q(1) - Q(2)) \\
\alpha_5 &= U(0) - U(3) & \beta_5 &= \frac{\sqrt{3}}{2} P(2) \\
\alpha_6 &= U(1) - U(2) \\
\alpha_7 &= V(0) - V(2)
\end{aligned}
$$

Then

$$
\begin{aligned}
A_0 &= \tfrac{1}{6}(\alpha_2 + \alpha_3) \\
A_1 &= \tfrac{1}{6}(\alpha_1 + \alpha_4) & B_1 &= \tfrac{1}{6}(\beta_1 + \beta_5) \\
A_2 &= \tfrac{1}{6}\left(\alpha_5 + \frac{\alpha_6}{2}\right) & B_2 &= \tfrac{1}{6}\beta_3 \\
A_3 &= \tfrac{1}{6}\alpha_7 & B_3 &= \tfrac{1}{6}\beta_2 \\
A_4 &= \tfrac{1}{6}\left(\alpha_2 - \frac{\alpha_3}{2}\right) & B_4 &= \tfrac{1}{6}\beta_4 \\
A_5 &= \tfrac{1}{6}(\alpha_1 - \alpha_4) & B_5 &= \tfrac{1}{6}(\beta_1 - \beta_5) \\
A_6 &= \tfrac{1}{6}(\alpha_5 - \alpha_6)
\end{aligned}
$$

These coefficients are associated with the Fourier series

$$F(x) = \frac{A_0}{2} + \sum_{k=1}^{n-1} \left(A_k \cos \frac{2\pi kx}{L} + B_k \sin \frac{2\pi kx}{L} \right) + \frac{A_n}{2} \cos \frac{2\pi Nx}{L}$$

5-6 SUMMARY

A number of important observations have been made on the Fourier expansion of discrete functions. Take, for example, the 12-point Fourier coefficient formulas. The 12 values of the discrete function result in a discrete spectrum with only six frequency components. In general, $2N$ values of a discrete function will result in a spectrum with only N spectral components. This generally holds true, reflecting the rule of thumb that data must be sampled at at least twice the highest frequency of interest for the coefficients of the spectral components to be determined at the frequency of interest.

The second observation deals with the physical characteristics of sampled data. Note that once a function is sampled the spectrum of the sequence of sampled functions has no frequency components greater than half the sampling frequency. What, then, happens to the high frequency components of a sampled continuous function? They are folded down into the low-frequency components of the discrete spectrum. They are summed with the low-frequency components. In this sense the high-frequency components are not actually rejected by the sampling process, but they are folded into the low-frequency components of the discrete function spectrum, resulting in distortion of the low-frequency components that made up the original continuous-function frequency spectrum. Thus, although the sampling operation does not result in high-frequency components in the sampled-functions spectrum, the true continuous-function spectrum can be much distorted by the sampling process in the low frequencies.

5-7 REFERENCES

For this chapter refer to Richard Hamming's *Numerical Methods for Scientists and Engineers* (McGraw-Hill, 1973), Chapter 1 and Chapters 31 through 34.

Example 5-1 Compute the coefficients in the Fourier series expansions of a continuous periodic triangular wave function. Then use these coefficients to compute the discrete function spectrum by way of equations 5-9 through 5-11.

The continuous periodic triangular wave form is shown in Figure 5-1. It is apparent that

$$f(t) = -f(-t)$$

and

$$f\left(t \pm \frac{T}{2}\right) = -f(t)$$

The first equation shows that $f(t)$ is an odd function and thus only the sine components are involved in the Fourier series approximation of $f(t)$. The second equation shows that only the odd harmonics are involved in the series. Furthermore, when two symmetry conditions exist, it is necessary to integrate only over one-quarter of the period of the function to determine the Fourier coefficients (an interesting property for the reader to demonstrate to himself). It follows, that

$$b_n = \frac{8}{T} \int_0^{T/4} f(t) \sin n\left(\frac{2\pi}{T}\right) t \, dt, \qquad (n, \text{odd})$$

Since

$$f(t) = \frac{4At}{T}, \qquad 0 \leqslant t \leqslant \frac{T}{4}$$

$$b_n = \frac{8}{T} \int_0^{T/4} \left[\left(\frac{4A}{T}\right) t \sin\left\{ n\left(\frac{2\pi}{T}\right) t \right\} \right] dt$$

$$b_n = \frac{8A}{n^2 \pi^2} \sin\left(\frac{n\pi}{2}\right), \qquad (n, \text{odd})$$

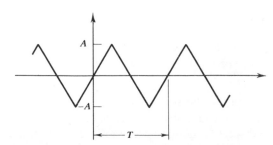

Figure 5-1   Continuous periodic triangular waveform.

Thus

$$b_n = \begin{cases} \dfrac{8A}{n^2\pi^2}, & n = 1,5,9,\cdots \\[4mm] \dfrac{-8A}{n^2\pi^2}, & n = 3,7,11,\cdots \end{cases}$$

The first 15 continuous function spectrum components are tabulated below.

$$b_1 = \frac{8A}{\pi^2} \qquad b_6 = 0 \qquad b_{11} = \frac{-8A}{121\pi^2}$$

$$b_2 = 0 \qquad b_7 = \frac{-8A}{49\pi^2} \qquad b_{12} = 0$$

$$b_3 = \frac{-8A}{9\pi^2} \qquad b_8 = 0 \qquad b_{13} = \frac{8A}{169\pi^2}$$

$$b_4 = 0 \qquad b_9 = \frac{8A}{81\pi^2} \qquad b_{14} = 0$$

$$b_5 = \frac{8A}{25\pi^2} \qquad b_{10} = 0 \qquad b_{15} = \frac{-8A}{225\pi^2}$$

The discrete spectrum components are developed by using equation 5-10 for the case $2N = 10$:

$$B_1 = b_1 + (b_{11} - b_9) + (b_{21} - b_{19}) + (b_{31} - b_{29}) + \cdots$$

$$B_2 - b_2 + (b_{12} - b_8) + (b_{22} - b_{18}) + (b_{32} - b_{28}) + \cdots$$

$$B_3 = b_3 + (b_{13} - b_7) + (b_{23} - b_{17}) + (b_{33} - b_{27}) + \cdots$$

$$B_4 = b_4 + (b_{14} - b_6) + (b_{24} - b_{16}) + (b_{34} - b_{26}) + \cdots$$

From the table of continuous function spectrum components we see that

$$B_1 \cong \frac{8A}{\pi^2}\left[\left(1 - \frac{1}{121}\right) - \left(\frac{1}{81}\right)\right] = 0.9793900\left(\frac{8A}{\pi^2}\right)$$

$$B_2 \cong \frac{8A}{\pi^2}[(0+0) - 0] = 0$$

$$B_3 \cong \frac{8A}{\pi^2}\left[\left(-\frac{1}{9} + \frac{1}{169}\right) - \left(\frac{-1}{99}\right)\right] = 0.0847859\left(\frac{8A}{\pi^2}\right)$$

$$B_4 \cong \frac{8A}{\pi^2}[(0+0) - (0)] = 0$$

The following table shows the discrete function spectrum elements and the continuous function spectrum elements side by side for easy comparison. The difference is the result of the aliasing phenomenon.

Continuous Function Spectrum Elements	Discrete Function Spectrum Elements
$b_1 = 1.0\left(\dfrac{8A}{\pi^2}\right)$	$B_1 = 0.9773900\left(\dfrac{8A}{\pi^2}\right)$
$b_2 = 0$	$B_2 = 0$
$b_3 = -0.1111111\left(\dfrac{8A}{\pi^2}\right)$	$B_3 = -0.0847659\left(\dfrac{8A}{\pi^2}\right)$
$b_4 = 0$	$B_4 = 0$

Example 5-2   Compute the Fourier coefficients for the Fourier series approximation of the discrete function

$$y_n = \sin(n\omega T)$$

where $\omega = 1$ Hz. This 12-point discrete function is tabulated below.

n (seconds)	nT (seconds)	$n\omega T$ (degrees)	$\mathrm{Sin}(n\omega T) = F(n)$
0	0	0	0.00
1	0.1	36	0.59
2	0.2	72	0.95
3	0.3	108	0.95
4	0.4	144	0.59
5	0.5	180	0.00
6	0.6	216	−0.59
7	0.7	252	−0.95
8	0.8	288	−0.95
9	0.9	324	−0.59
10	1.0	360	0.00
11	1.1	396	+0.59

A closer examination of the table reveals that the sine function is tabulated over the interval to 1.2 seconds, while the function is periodic on the interval 1 second. Clearly, the coefficients that we generate using the

12-point formulas apply to Fourier series approximations of the tabulated functions exactly as shown (with a period of 1.2 seconds, not 1 second), and not the coefficients for the Fourier series approximation of a pure 1 Hz sine wave. The purpose for selecting this unusual problem is that it illustrates not only the use of the 12-point formula, but also the effect of one of the practical problems associated with sampling periodic functions. The reality of sampling functions from experiments is that the functions are often not exactly periodic or, if they are, the period is not known precisely and some approximation of the period must be made. This example might be considered the result of an experiment where an estimate of the period of the function being sampled is made to be 1.2 seconds, where in reality the periodic function repeats on the interval 1 second. Following the 12-point Fourier analysis procedure, the discrete function 12-point Fourier coefficients are generated from the arrays as tabulated in Table 5-1. The array numbers in the table correspond to the array numbers in the text and are shown here for the sake of convenience. The numerical evaluation of the 12-point Fourier series coefficients are summarized in

Table 5-1 Discrete Function 12-point Fourier Series Coefficient Generation Arrays

$(0) \rightarrow F(6)$	0.00	0.59	0.95	0.95	0.59	0.00	-0.59		(*A*1)
$(12) \leftarrow F(7)$		0.59	0.00	-0.59	-0.95	-0.95			
dd	0.00	1.18	0.95	0.36	-0.36	-0.95	-0.59		(*A*2)
ubtract		0.00	0.95	1.54	1.54	0.95			
	$S(0)$	$S(1)$	$S(2)$	$S(3)$	$S(4)$	$S(5)$	$S(6)$		(*A*2)
		$T(1)$	T(2)	$T(3)$	$T(4)$	$T(5)$			
$(0) \rightarrow S(3)$	0.00	1.18	0.95	0.36	0.00	0.95	1.54	$T(1) \rightarrow T(3)$	(*A*3)
$(6) \leftarrow S(4)$	-0.59	-0.95	-0.36		0.95	1.54		$T(5) \leftarrow T(4)$	
dd	-0.59	0.23	0.59	0.36	0.95	2.49	1.54		(*A*4)
ubtract	0.59	2.13	1.31		-0.95	-0.59			
	$(U(0)$	$U(1)$	$U(2)$	$U(3)$	$P(1)$	$P(2)$	$P(3)$		(*A*4)
	$V(0)$	$V(1)$	$V(2)$		$Q(1)$	$Q(2)$			

Table 5-2. The A-coefficients are associated with the cosine components and the B-coefficients are associated with the sine components. Also, the table shows a check of the initial conditions. At $t=0$, the discrete function starts at 0. Therefore, the sum of the cosine amplitude coefficients should equal 0, as they do.

Though we could discuss the tail effects by examining the individual elements of the series expansion, it is more convenient to use the power spectrum or amplitude spectrum as a means of discussing this phenomenon. The 12-point spectrum calculations are tabulated in Table 5-3. First note that the DC component of the spectrum is given by P_0. This indicates that the average effect of the "tail" of our irregular periodic coefficient is to bias the otherwise 0-DC coefficient to the level 0.049. Second, the lowest-frequency component (the fundamental frequency equal to $1/1.2$ equal to $0.83333...$) contains the greatest amount of power of all of the harmonics. Clearly, this is so because it is the closest frequency to the 1 Hz periodic function that we have sampled. The power in the next-highest harmonic is approximately one-tenth that of the fundamental.

Table 5-2 12-Point Fourier Series Coefficient Calculations

Cosine components

$$2A_0 = \frac{1}{6}(-0.59+0.36+0.23+0.59)=0.098$$

$$A_1 = \frac{1}{6}(0.59+0.655+0.866\times2.13)=0.5149$$

$$A_2 = \frac{1}{6}(-0.59-0.36-.18)=-0.188$$

$$A_3 = \frac{1}{6}(-.72)=-0.12$$

$$A_4 = \frac{1}{6}(0.59+0.36-.41)=-0.107$$

$$A_5 = \frac{1}{6}(0.59+.655-0.866\times2.13)=-0.0999$$

$$2A_6 = \frac{1}{6}(-0.59-0.36-0.23+0.59)=-0.098$$

Note that at $t=0$, $\Sigma Ai \equiv 0$, as it should

Sine components

$$B_1 = \frac{1}{6}(0.475+1.54+0.866\times2.489)=0.695$$

$$B_2 = \frac{0.866}{6}(-0.95-0.59)=-0.2223$$

$$B_3 = \frac{1}{6}(0.95-1.54)=0.098$$

$$B_4 = \frac{0.866}{6}(-0.95+0.59)=-0.05196$$

$$B_5 = \frac{1}{6}(0.475+1.54-0.866\times2.49)=-0.0436$$

Table 5-3 12-Point Spectrum Calculations

ω Hz	Power Spectrum (rounded)	Amplitude Spectrum
0	$P_0 = A_0^2/4\ \ =0.00$	$\sqrt{P_0}\ \ =0.049$
0.833	$P_1 = A_1^2 + B_1^2 = 0.75$	$\sqrt{P_1}\ \ =0.865$
1.666	$P_2 = A_2^2 + B_2^2 = 0.08$	$\sqrt{P_2}\ \ =0.291$
2.499	$P_3 = A_3^2 + B_3^2 = 0.02$	$\sqrt{P_3}\ \ =0.155$
3.333	$P_4 = A_4^2 + B_4^2 = 0.01$	$\sqrt{P_4}\ \ =0.119$
4.166	$P_5 = A_5^2 + B_5^2 = 0.01$	$\sqrt{P_5}\ \ =0.109$
5.499	$P_6 = A_6^2/4\ \ =0.00$	$\sqrt{P_6}\ \ =0.049$

Had we taken the 12 sample points equally distributed over the periodic sine wave function, we would have found a single harmonic component at 1 Hz and the rest of the components would have been zero or very small, depending only on truncation error as related to the number of terms carried in the pocket calculator analysis. Here we see the effect of the "tail" is to affect the DC level and spread the power in the 1 Hz sine function over higher-frequency harmonics. The reason for this is that the high-frequency components are required to take care of the discontinuous end effects associated with the "tail" in our sampled periodic discrete function. Specifically, this "tail" is associated with the jump discontinuity in going from $+0.59$ at $n=11$ to 0 at $n=12$ for the example function that we have chosen to analyze. Hopefully this example will interest the students who read this book in further readings in practical Fourier analysis, on which there is an extensive literature.

CHAPTER 6

NUMERICAL INTEGRATION

6-1 INTRODUCTION

There are basically two types of integral with which we are concerned in this chapter: the definite integral and the indefinite integral. The definite integral is given by the formula

$$y(b) = y(a) + \int_a^b f(x)\,dx \qquad (6\text{-}1)$$

and the indefinite integral is defined by

$$y(x) = y(a) + \int_a^x f(t)\,dt \qquad (6\text{-}2)$$

The definite integral is characterized by computing the area under the curve of a bounded function; the indefinite integral can be thought of as computing the antiderivative of the integrand and thus generating the sequence of values of a function. We study definite integrals from the standpoint of quadrature—that is, for computing the area under a curve. We study indefinite integrals from the standpoint of integrating differential equations. Our first concern here is the definite integral.

6-2 DEFINITE INTEGRATION

Computing the area under an arbitrary curve is usually based on the concept of analytic substitution. The idea is to use a known function whose definite integral is easily evaluated to substitute for the arbitrary function to be integrated. The integration is actually performed on the

154

substitute function and attributed to the integral of the arbitrary function to the degree to which it approximates the latter. In classical mathematics the substitute functions to be integrated are usually polynomials. The polynomial is then analytically integrated and, insofar as the polynomial approximates the continuous function, the integral is attributed to the integral of the arbitrary function. When the integrand is a polynomial of degree n and the approximating function is also a polynomial of degree n, the formula can be made exact by appropriately selecting the coefficients in the integration formula.

The process of *analytic substitution* or of other means of approximating definite and indefinite integrals is so fascinating that virtually every numerical analyst finds new ways to rederive many of the classical formulas and a few others as well. Though one is tempted to present the most sophisticated integration methods, the focus here remains on classical developments, which are straightforward and easy to apply on the pocket calculator. The reader should be aware, however, of the tremendous quantity of good mathematics in numerical integration developed in the last 20 years. This is due to numerical calculations being done on the digital computer and to the use of numerical analysis in sophisticated technology problems in varied areas. Structures, communications systems, control systems, design of aircraft, and the design of chemical plants are areas where the simulation of systems with widely separated eigenvalues and the numerical integration of functions that are almost neutrally stable (at large integration step size), have produced new integration concepts based on the technology to which they were being applied. Structural dynamicists have developed special numerical integration formulas for integrating their "stiff differential equations." Controls analysts have produced such formulas based solely on frequency-domain considerations. And special single-step real-time numerical integration formulas have been developed by simulation scientists.

These problems can be encountered on the pocket calculator, especially the programmable pocket calculator. Here, however, we focus on the more classical formulas, which have fairly general and broad applications to the more analytically tractable functions. Furthermore, there is a vast body of literature on these classical methods for further reference, should it be required.

Trapezoidal Integration

If we approximate the function $f(x)$ on a bounded interval $a \leqslant x \leqslant b$ by a line through the end points, we can write the equation for the approximating function over the interval as

$$y(x) = f(a) + \left[\frac{f(b) - f(a)}{b - a}\right](x - a)$$

$$(6\text{-}3)$$

$$y(x) = \frac{(b - x)f(a) + (x - a)f(b)}{b - a}$$

Integrating equation 6-3, we find

$$\int_a^b y(x)\,dx = \left(\frac{f(b) + f(a)}{2}\right)(b - a) \qquad (6\text{-}4)$$

Equation 6-4 computes the area under the straight-line interpolation between the two end points. This is called trapezoidal integration because
this area is enclosed by a trapezoid formed by lines connecting the end
points, the abscissa, and the vertical lines connecting the end points to the
abscissa. If the interval is large, the trapezoidal approximation can lead to
large numerical integration error. This is resolved by a repeated application of the trapezoidal rule on smaller intervals of the dependent variable.
When this is done for equally spaced intervals, Δx, trapezoidal integration
takes the form

$$\int_a^b f(x)\,dx = \Delta x\left(\frac{f(a)}{2} + f(a + \Delta x) + f(a + 2\Delta k) + \cdots + \frac{f(b)}{2}\right) \qquad (6\text{-}5)$$

Trapezoidal integration, though not the simplest one to derive or compute (Euler, modified Euler, or rectangular integration are simpler concepts) and its error formula does not give the least error for the fewest
computations, is straightforward to apply on the pocket calculator and is
easily remembered. As we move to integration formulas involving mid-
values and their derivatives, estimates of a roundoff and truncation error,
and adjustments of phase shift and amplitude, we retreat further from
simple visualizations of the integration process and must increasingly rely
on the rationale for their development to be assured of their applicability
to a problem. Ultimately, analytical integration is compared with the
approximate numerical integration to evaluate the difference between
several methods of integration for a particular problem. Clearly, this is an
overkill for back-of-the-envelope engineering analysis or analysis on the
pocket calculator intended simply to compute the area under the curve of a
given function. If trapezoidal integration is sufficiently accurate, and the
number of intervals needed to obtain the desired accuracy is not prohibi-
tive, it is very useful for pocket calculator analysis.

6-3 ERROR IN TRAPEZOIDAL INTEGRATION

We do not here aim to explore the derivation of integration or error formulas—merely to tabulate the commonly used ones and put them in a form that is immediately useful for the pocket calculator. Nevertheless, it is instructive to examine the error of a simple integration formula as a means for understanding the error equations given for the more sophisticated integration formulas. Following Hamming, then, we examine the truncation error in the trapezoidal integration algorithm by substituting a Taylor series expansion into the integration formula. By comparing both sides of the results, we can then determine the error associated with the analytic substitution process in the numerical integration. Specifically, if we write the integrand in its Taylor series expanded form as

$$f(x) = f(a) + (x-a)f'(a) + \frac{(x-a)^2}{2!}f''(a) + \cdots \qquad (6\text{-}6)$$

and substitute this into both sides of the trapezoidal integration formula, we find that, on integration, the left side becomes

$$\frac{(b-a)}{1!}f(a) + \frac{(b-a)^2}{2!}f'(a) + \frac{(b-a)^3}{3!}f''(a) + \cdots \qquad (6\text{-}7)$$

The right side becomes

$$\frac{\Delta x}{2}\left[f(a) + (b-a)f'(a) + \frac{(b-a)^2}{2}f''(a) + \cdots + f(a) \right] + \epsilon \qquad (6\text{-}8)$$

where $\Delta x = (b-a)$. After canceling like terms on both sides we can derive the truncation error formula:

$$\epsilon + \frac{(b-a)^3}{4}f''(a) + \cdots = \frac{(b-a)^3}{3!}f''(a) \qquad (6\text{-}9)$$

$$\epsilon = \left(\frac{1}{3!} - \frac{1}{4} \right)(b-a)^3 f''(a) - \frac{(b-a)^4}{5}f'''(a) - \cdots \qquad (6\text{-}10)$$

If we assume the largest part of the error term to be given by the first term in its series expansion, we can write

$$\epsilon \approx - \frac{(b-a)^3 f''(a)}{12} \qquad (6\text{-}11)$$

or, more generally,

$$\epsilon \approx - \frac{(b-a)^3 f''(\theta)}{12} , \qquad (a \leqslant \theta \leqslant b) \qquad (6\text{-}12)$$

If, however, the function has contributions to the error formula that are large in the higher-order terms, this error formula does not apply. It is applicable for many of the practical engineering problems, and thus is generally quoted as the error associated with trapezoidal integration.

The specific error formula for trapezoidal integration is less important here than is the method by which it is derived. We used the Taylor series expansion for the integrand in order to derive a Taylor series truncation term for "the area." Another alternative would have been to use a Fourier series representation of the function to determine the truncation in the frequency domain. Another approximating polynomial could have been the Chebyshev polynomial approximation of $f(x)$, which would have given another type of truncated polynomial approximation error formula. While the interpretation of the results of each error formula is different, the magnitude of the error is not. The error is a characteristic of the integration formula, rather than the approximating polynomial used in the error formula evaluation.

Figure 6-1 shows that for concave-up type of functions trapezoidal integration is always slightly more than the curve it is trying to approximate; for concave-down type of functions it is slightly less. Thus it seems

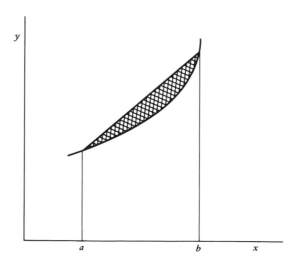

Figure 6-1   Truncation error in trapezoidal integration.

reasonable to expect, when integrating "wavy" functions, the intervals to be set up so that, at a minimum, the eyeball approximation of the errors on one interval may have a chance to cancel the errors on the other interval. We can extend the error formula for simple trapezoidal integration to the composite formula by similar reasoning:

$$\epsilon \approx - \frac{(b-a)\Delta x^2}{12} f''(\theta), \qquad (a \leqslant \theta \leqslant b) \tag{6-13}$$

Writing error formulas such as equations 6-12 and 6-13 is, of course, easier than evaluating them meaningfully. One approach is to find the second derivative of the function being considered, compute the minimum error and the maximum error, and divide by 2 to obtain the average error of the integration over the interval. Another approach is to take the worst-case error. A great number of other alternatives also exist. The question is, what is the criterion for numerically evaluating the error? Unfortunately, there is no easy answer to this question. From an engineering viewpoint, the error defined by equation 6-9 perhaps has more meaning than those most often quoted in numerical analysis books. In this sense the process of deriving the error formula is the more fundamental issue in that the engineer or scientist can compute his own error formula suited to his specific problem.

Another aspect of the numerical error formulas associated with integration formulas is that they are absolute errors, whereas the error of interest is usually relative error. Again, the author has no easy solution of the problem of deriving relative error formulas for numerical analysis. The difficulty is pointed out here to warn the student or first-time numerical analyst about error formulas in general. Preferably he should derive his own formula for a particular problem being numerically analyzed. An estimate of the error in a numerical approximation over an analytical calculation must be made, but its interpretation is not straightforward and the result cannot be casually given from questionable error formulas.

6-4 MIDPOINT INTEGRATION

Midpoint integration uses the midvalue of an interval and the derivative of the integrand evaluated at the midvalue to define the slope at the midpoint of the interval, again forming a trapezoid whose area under the curve approximates that of a function to be considered.

The midpoint integration formula as developed by Hamming is easy to follow and nicely introduces the concept of a general approach to deriving polynomial approximations for analytic substitution. We are to derive an integration formula of the form

$$\int_a^b f(x)\,dx = w_1 f\left(\frac{a+b}{2}\right) + w_2 f'\left(\frac{a+b}{2}\right) \qquad (6\text{-}14)$$

Hamming's weighting coefficients can be easily derived by noting that we first require this formula to be exact for $f(x) = 1$. This gives

$$b - a = w_1 \qquad (6\text{-}15)$$

We also require that this formula be exact for $f(x) = x$, which leads to

$$\frac{b^2 - a^2}{2} = w_1\left(\frac{a+b}{2}\right) + w_2 \qquad (6\text{-}16)$$

We can determine the two Hamming's coefficients by solving these equations simultaneously:

$$w_2 = \left(\frac{b^2 - a^2}{2}\right) - \frac{(b-a)(a+b)}{2} = \left(\frac{b^2 - a^2}{2}\right) - \left(\frac{b^2 - a^2}{2}\right) = 0$$

$$w_1 = b - a \qquad (6\text{-}17)$$

We therefore find the midvalue integration formula to be

$$\int_a^b f(x)\,dx = (b - a)f\left(\frac{a+b}{2}\right) \qquad (6\text{-}18)$$

We see that midpoint integration developed in this manner results in rectangular integration. That is, the area formed by the rectangle sampled at the midvalue is identically equal to the area under the tangent line at the midvalue of the interval. At first it might seem paradoxical that the low-order rectangular integration could be as good as trapezoidal integration—that formulas based on a single point of f could be as accurate as a two-point trapezoidal formula. In fact, rectangular integration can be made as precise as desired if the sample point on a bounded interval can be varied until the mean value theorem of calculus is satisfied. *Once again*, rectangular integration can be made as precise as the true integral provided that the point at which the function is sampled on the interval can be determined, so that the rectangle formed by the sampled value and the lines connecting the end points of the function on the interval and the abscissa itself have the same area as that under the function bounded on the interval. This fact is reflected by equation 6-18.

Again, to find the truncation error term, we use the Taylor series:

$$f(x) = f(a) + \frac{(x-a)}{1!}f'(a) + \frac{(x-a)^2}{2!}f''(a) + \cdots \qquad (6\text{-}19)$$

where, upon substituting in both sides of equation 6-18, we find

$$\epsilon + \frac{(b-a)^3}{8} + \cdots = \frac{(b-a)^3}{6} \qquad (6\text{-}20)$$

This is usually simplified to

$$\epsilon \approx \frac{(b-a)^3 f''(a)}{20} \qquad (6\text{-}21)$$

or, more generally, for $a \leqslant \theta \leqslant b$

$$\epsilon \approx \frac{(b-a)^3 f''(\theta)}{20}$$

Comparing equations 6-21 and 6-12, we see that midpoint rectangular integration is more accurate than endpoint trapezoidal integration even though the rectangular integration is based on knowing the function at only one point while the trapezoidal rule ´of integration requires the knowledge of the function at two end points.

Extending the midpoint integration formula, we find, as in the trapezoidal formula, the composite midpoint integration formula to be of the form

$$\int_a^b f(x)\,dx$$

$$= \Delta x\left[f\left(a + \frac{\Delta x}{2}\right) + f\left(a + \frac{3\Delta x}{2}\right) + f\left(a + \frac{5\Delta x}{2}\right) + \cdots + f\left(b - \frac{\Delta x}{2}\right) \right] + \epsilon$$

where its error formula is given by

$$\epsilon \approx \frac{(b-a)\Delta x^2}{24}f''(\theta), \qquad (a \leqslant \theta \leqslant b) \qquad (6\text{-}22)$$

Note also that extended trapezoidal integration can be modified to include end points outside the interval $[a,b]$. The modified trapezoidal rule is given by

$$\int_a^b f(x)\,dx = \Delta x \left[\frac{f(a)}{2} + f(a+\Delta x) + f(a+2\Delta x) + \cdots + \frac{f(b)}{2} \right]$$

$$+ \frac{\Delta x}{24} [-f(a-\Delta x) + f(a+\Delta x) + f(b-\Delta x) - f(b+\Delta x)] \quad (6\text{-}23)$$

where the error associated with modified trapezoidal integration is given by

$$\epsilon = \frac{11(b-a)\Delta x^4}{720} f''''(\theta), \qquad (a+\Delta x) \leqslant \theta \leqslant (b+\Delta x) \qquad (6\text{-}24)$$

which is usually much more accurate than extended midpoint integration with only slightly more work.

Other Popular Definite Integration Formulas

Simpson's rule, perhaps the most commonly used integration formula, is given by

$$\int_0^{2\Delta x} f(x)\,dx = \frac{\Delta x}{3}(f_0 + 4f_1 + f_2) \qquad (6\text{-}25)$$

Its associated error formula is given by

$$\epsilon = -\frac{\Delta x^5}{90} f''''(\theta), \qquad (0 \leqslant \theta \leqslant 2\Delta x)$$

Simpson's rule has the nice property that it integrates cubics exactly even though it samples only three points of the integrand and in addition has very small error terms when Δx is less than 1 and on the order of one-half.

Simpson's rule can also be extended (on an even number of intervals) according to the formula

$$\int_{x_0}^{x_{2n}} f(x)\,dx = \frac{\Delta x}{3}(f_0 + 4f_1 + 2f_2 + 4f_3 + 2f_4 + \cdots + f_{2n}) \qquad (6\text{-}26)$$

Its error formula is given by

$$\epsilon = \frac{n\Delta x^5}{90} f''''(\theta), \qquad (x_0 \leqslant \theta \leqslant x_0 + 2n\Delta x) \qquad (6\text{-}27)$$

Perhaps the simplest extended integration formula is the Euler-Maclaurin formula:

$$\int_{x_0}^{x_n} f(x)\,dx = \Delta x\left(\frac{f_0}{2}+f_1+f_2+f_3+\cdots+\frac{f_n}{2}\right)-\left(\frac{B_2\Delta x^2}{2!}\right)(f'_n-f'_0)-\cdots$$

$$-\left(\frac{B_{2k}\Delta x^{2k}}{2k!}\right)(f_n^{(2k-1)}-f_0^{(2k-1)})+\epsilon_{2k} \qquad (6\text{-}28)$$

It has the error formula

$$\epsilon_{2k}=\left\{\frac{\theta n B_{2k+2}\Delta x^{(2k+3)}}{(2k+2)!}\right\}\left\{\max_{x_0\leqslant x\leqslant x_n}|f(x)^{2k+2}|\right\}, \qquad (-1\leqslant\theta\leqslant 1) \quad (6\text{-}29)$$

Here B_{2k} is a Bernoulli number.

The three-eights rule for definite integration is given by

$$\int_{x_0}^{x_3} f(x)\,dx=\frac{3\Delta x}{8}(f_0+3f_1+3f_2+f_3) \qquad (6\text{-}30)$$

Its associated error formula is

$$\epsilon=-\frac{3\Delta x^5}{80}f''''(\theta), \qquad x_0\leqslant\theta\leqslant x_3 \qquad (6\text{-}31)$$

Two types of formulas are used for quadrature when many sample points of the integrand are known: Bode's definite integral formulas and the Newton-Cotes formulas of the open type. Bode's rules for quadrature are shown in Table 6-1, and the Newton-Cotes formulas are tabulated in Table 6-2.

The high-order formulas, such as the Newton-Cotes and Bode's formulas, can have some very undesirable properties for large n. For some analytic and discrete functions the sequence of the integrals of the interpolating polynomials does not converge toward the integral of the function. Also, the coefficients in these integration formulas are large and of alternating sign, which is undesirable from the standpoint of propagating roundoff error. It is primarily for these reasons that the Newton-Cotes formulas are rarely used for high values of n. For lower values of n they can be simplified to some other well-known formula, such as the previously discussed trapezoidal formula and Simpson's rule. Although Bode's rule gets around the alternating signs associated with the Newton-Cotes formulas, it too has convergence problems for certain occasionally encountered functions. Suffice it to say that the extended trapezoidal integration with

Table 6-1 Bode's Definite Intergration Formulas for Integrating Functions Whose End Points Are Known

Integration Formulas	Error Formulas
$\displaystyle\int_{x_0}^{x_4} f(x)\,dx = \frac{2\Delta x}{45}(7f_0 + 32f_1 + 12f_2 + 32f_3 + 7f_4)$	$-\dfrac{8\Delta x^7 f^{VI}(\theta)}{945}$
$\displaystyle\int_{x_0}^{x_5} f(x)\,dx = \frac{5\Delta x}{288}(19f_0 + 75f_1 + 50f_2 + 50f_3 + 75f_4 + 19f_5)$	$-\dfrac{275\Delta x^7 f^{VI}(\theta)}{12096}$
$\displaystyle\int_{x_0}^{x_6} f(x)\,dx = \frac{\Delta x}{140}(41f_0 + 216f_1 + 27f_2 + 272f_3 + 27f_4 + 216f_5 + 41f_6)$	$-\dfrac{9\Delta x^9 f^{VIII}(\theta)}{1400}$
$\displaystyle\int_{x_0}^{x_7} f(x)\,dx = \frac{7\Delta x}{17280}(751f_0 + 3577f_1 + 1323f_2 + 2989f_3 + 2989f_4$ $+ 1323f_5 + 3577f_6 + 751f_7)$	$-\dfrac{8183\,\Delta x^9 f^{VIII}_{(\theta)}}{518400}$
$\displaystyle\int_{x_0}^{x_8} f(x)\,dx = \frac{4\Delta x}{14175}(989f_0 + 5888f_1 - 928f_2 + 10496f_3 - 4540f_4$ $+ 10496f_5 - 928f_6 + 5888f_7 + 989f_8)$	$-\dfrac{2368\,\Delta x'' f^X_{(\theta)}}{467775}$

end effect modification has high accuracy, does not propagate roundoff, requires only a reasonable amount of work in computing the integral of any function, and is thus recommended for analysis on the pocket calculator.

6-5 INDEFINITE NUMERICAL INTEGRATION

Indefinite numerical integration is the numerical method for solving differential equations. Given the equation

$$\frac{dy}{dx} = f(x,y) \tag{6-32}$$

we would usually solve it by indefinite integration as follows:

$$y = y_0 + \int_{x_0}^{x} f(t,y)\,dt \tag{6-33}$$

Table 6-2 Newton-Cotes' Definite Integration Formulas for Integrating Functions Whose End Points Are Undefined or Unknown or Are Singular Points

Integration Formulas	Error Formulas
$\displaystyle\int_{x_0}^{x_3} f(x)\,dx = \frac{3\Delta x}{2}(f_1 + f_2)$	$\dfrac{\Delta x^3}{4} f^{\mathrm{II}}(\theta)$
$\displaystyle\int_{x_0}^{x_4} f(x)\,dx = \frac{4\Delta x}{3}(2f_1 - f_2 + 2f_3)$	$\dfrac{28\Delta x^5}{90} f^{\mathrm{IV}}(\theta)$
$\displaystyle\int_{x_0}^{x_5} f(x)\,dx = \frac{5\Delta x}{24}(11f_1 + f_2 + f_3 + 11f_4)$	$\dfrac{95\Delta x^5}{144} f^{\mathrm{IV}}(\theta)$
$\displaystyle\int_{x_0}^{x_6} f(x)\,dx = \frac{6\Delta x}{20}(11f_1 - 14f_2 + 26f_3$ $\qquad\qquad - 14f_4 + 11f_5)$	$\dfrac{41\Delta x^7}{140} f^{\mathrm{VI}}(\theta)$
$\displaystyle\int_{x_0}^{x_7} f(x)\,dx = \frac{7\Delta x}{1440}(611f_1 - 453f_2 + 562f_3$ $\qquad\qquad + 562f_4 - 453f_5 + 611f_6)$	$\dfrac{5257\Delta x^7}{8640} f^{\mathrm{VI}}_{(\theta)}$
$\displaystyle\int_{x_0}^{x_8} f(x)\,dx = \frac{8\Delta x}{945}(460f_1 - 954f_2 + 2196f_3 - 2459f_4$ $\qquad\qquad + 2196f_5 - 954f_6 + 460f_7)$	$\dfrac{3956\Delta x^9}{14175} f^{\mathrm{VII}}_{(\theta)}$

It is apparent that the solution of the differential equation depends on its own evaluation of the integral. This is precisely the chief problem in indefinite integration; that is, indefinite integrals are in an implicit form.

Note that an explicit indefinite integral takes the form

$$y = y_0 + \int_{x_0}^{x} f(t)\,dt \qquad\qquad (6\text{-}34)$$

which is a special case of the differential equation

$$\frac{dy}{dx} = f(x) \qquad\qquad (6\text{-}35)$$

Clearly, this type of numerical integration can be performed analytically, hence is not of concern here.

The simplest indefinite numerical integration algorithm is Euler's integration formula:

$$y_{n+1} = y_n + \Delta x \left(\frac{dy}{dx} \right)_n \qquad (6\text{-}36)$$

Here we see that a new estimate (y_{n+1}) of y is based on the old estimate (y_n) and its derivative $[(dy/dx)_n]$. The derivative is usually calculated directly from the differential equation once y is estimated. Since the new estimate y_{n+1} is based on the old estimate y_n' and the old value y_n it is clearly an "open-loop" process where the new value y_n is based on an extrapolation from previously known data and thus is subject to extrapolation errors. The process of determining new values of y is really a simple extension of determining the *direction field* associated with a solution of a differential equation. In general, the approach is to start at some initial condition (x_0, y_0) and calculate the slope, using the differential equation:

$$y_0' = f(f_0, y_0)$$

One then moves an interval Δx in the direction of the slope to a second point, which we now regard as the new initial point, and repeat the process iteratively. If small enough steps are taken we can reasonably hope that the sequence of solution values given by this procedure will lie close to the solution of the differential equation. In general, all of the elements of solving differential equations using numerical indefinite integration are present here. A table of the values of x, y, y', and Δy must be computed at each step in the numerical integration process. Also, the problem must be defined by specifying not only the differential equation and its initial conditions, but also the interval over which it is desired to solve the equation. It is then possible to select a convenient integration interval, and an integration formula that is accurate for that interval. For example,

$$\frac{dy}{dx} = e^{-y} - x^2$$

with initial conditions $y = 0$, $x = 0$. When integrated with Euler's integration formula

$$y_n = y_{n-1} + \Delta x y_{n-1}'$$

requires a specification of the interval Δx. The simplest approach is to experimentally determine the Δx that will accurately (as judged by the analyst) integrate the differential equation. Consider solutions of this differential equation with $\Delta x = 0.05, 0.1, 0.2,$ and 0.3. The results are tabulated in Table 6-3. A comparison of the numerically integrated solutions with the exact solution shows that the sensitivity of the solutions accuracy depends strongly on the integration step size. This is true, in general, for all numerical integrators when the integration step size is even a reasonable fraction of the "response time" of the differential equation.*

Table 6-3 Solution of $dy/dx = e^{-y} - x^2$

	Exact Solution	Euler Integrated Solution			
x	y	$\Delta x = 0.05$	$\Delta x = 0.10$	$\Delta x = 0.2$	$\Delta x = 0.3$
0.0	0.0				
0.1	0.09498	0.09694	0.09900	—	—
0.2	0.17977	0.18261	0.18557	0.19200	—
0.3	0.25389	0.25672	0.25964	—	0.27300
0.4	0.31667	0.31872	0.32077	0.32506	—
0.5	0.36731	0.36786	0.36833	—	—
0.6	0.40488	0.40329	0.40152	0.39756	0.39333
0.7	0.42839	0.42407	0.41942	—	—
0.8	0.43686	0.42923	0.42119	0.40395	—
0.9	0.42929	0.41782	0.40582	—	0.35277
1.0	0.40477	0.38895	0.37264	0.33749	—

A disadvantage of the Euler method is that it introduces systematic phase shift or lag (extrapolation) errors at each step. The procedure can be modified (modified Euler integration) to give better results—that is, greater accuracy for essentially the same method and the same amount of work.

6-6 THE MODIFIED EULER INDEFINITE INTEGRATION METHOD

An alternative to introducing lag into the calculation is to arrange the sampling so that the integrand is sampled not at the end point of the interval over which the integration is taking place but at the midpoint. This is similar to the development of the midpoint trapezoidal formula developed in Section 6-4. The task is to perform the integral

$$\int_{x_{n-1}}^{x_{n+1}} y'(x)\,dx \qquad (6\text{-}37)$$

*approximately the time required to move from one equilibrium condition to another using the midpoint formula (see Section 6-4). We wish to predict the next value of y based on present and past values of the independent variable. The midvalue prediction leads to

$$p_{n+1} = y_{n-1} + 2\Delta x y'_n \tag{6-38}$$

Using this predicted value, we can now compute the slope at the predicted solution point, by way of the differential equation,

$$p'_{n+1} = f(x_{n+1}, p_{n+1}) \tag{6-39}$$

and then apply the trapezoidal rule developed previously to update the estimate of the predicted solution point:

$$y_{n+1} = y_n + \frac{\Delta x}{2}(p'_{n+1} + y'_n) \tag{6-40}$$

The correction is called the corrected value of y_{n+1}. It is apparent that we are using the average of the slopes at the two end points of the interval of integration as the average slope in the interval.

In summary, this method has three steps:

Step 1 Predict the value of y_{n+1}, given the formula

$$p_{n+1} = y_{n-1} + 2\Delta x y'_n \tag{6-41}$$

Step 2 Compute the derivative at the predicted value, using the differential equation that describes the system:

$$p'_{n+1} = f(x_{n+1}, p_{n+1}) \tag{6-42}$$

Step 3 Make a second estimate of the value of y_{n+1}, using trapezoidal integration:

$$y_{n+1} = y_n + \frac{\Delta x}{2}(y'_n + y'_{n+1}) \tag{6-43}$$

This process of prediction and correction has led to the naming of this type of integration as the predict-correct concept of numerical integration. A number of predict-correct algorithms are tabulated at the end of this chapter; they can be used for indefinite integration of differential equations on the pocket calculator.

6-7 STARTING VALUES

In our previous analysis we assumed that we had values for the dependent and independent variables at the starting or initial point. However, the algorithm requires not only starting values, but also earlier values. The previous values can be obtained in two ways. They can be computed on the pocket calculator, or they can be analytically hand calculated. Both methods will be presented here.

The hand calculation method is based on the use of the Taylor series expansion of the function:

$$y(x+\Delta x) = y(x) + \Delta x y'(x) + \frac{\Delta x^2}{2} y''(x) + \cdots \tag{6-44}$$

The derivatives to be evaluated in the Taylor series expansion can be found from the differential equation by repeated differentiation. The number of terms of course depends on the step size and the accuracy desired. But, again, these are matters that can all be easily evaluated on the pocket calculator and the number of terms required can be empirically determined by continuing to take them until the desired accuracy is achieved.

The method for machine calculation is based on repeated use of the corrector formula. Again, if we are given the initial point (x_0, y_0), we can estimate the earlier point (x_{-1}, y_{-1}) by way of the "unmodified" Euler integration, working backwards as follows:

$$x_{-1} = x_0 - \Delta x$$

$$y_{-1} = y_0 - \Delta x y_0' \qquad \text{(first estimate of } y_{-1}) \tag{6-45}$$

We can use the estimate of the previous value of y combined with the differential equation to evaluate the derivative at the previous value of y. The trapezoidal corrector formula can then be repeated to iteratively correct the previous estimate until it achieves the accuracy desired for the calculation. The system of equations for the correction process become

$$y'_{-1} = f(x_{-1}, y_{-1}) \qquad \text{(first estimate of } y'_{-1})$$

$$y_{-1} = y_0 - \frac{\Delta x}{2}(y_0' + y'_{-1}) \qquad \text{(second estimate of } y_{-1}) \tag{6-46}$$

$$y'_{-1} = f(x-1, y_{-1}) \qquad \text{(second estimate of } y'_{-1})$$

$$\vdots \qquad\qquad\qquad\qquad \vdots$$

If, after a few iterations, the previous value of y does not stabilize, the integration step size can be halved, the previous value of $y_{n-1/2}$ computed, and the process repeated to compute y_{n-1}. Another alternative is to use the value of $y_{n-1/2}$ to estimate the value of $y_{n+1/2}$, the process repeated to take a half step forward to y_{n+1}, and then these values used as the starting values for the predict-correct integration algorithm.

6-8 ERROR ESTIMATES AND MODIFYING THE PREDICT-CORRECT PROCESS

The predictor formula just discussed is a midpoint integration formula that has the error equation

$$\epsilon_p = \frac{\Delta x^3}{3} y^{\text{III}}(\theta) \tag{6-47}$$

The corrector formula given in Section 6-3 has the error formula

$$\epsilon_c = -\frac{\Delta x^3}{12} y^{\text{III}}(\theta) \tag{6-48}$$

Since these error formulas are of opposite sign, the difference between the predicted value and the corrected value gives

$$y_p = y_c \approx (y_{\text{exact}} - \epsilon_p) - (y_{\text{exact}} - \epsilon_c) \tag{6-49}$$

Thus at any given step the difference between the predicted value and the corrected value is

$$-\frac{5}{12} \Delta x^3 y^{\text{III}}(\theta) \tag{6-50}$$

Furthermore, we see from equation 6-49 that approximately four-fifths of the difference results from the predictor component and one-fifth from the corrector component. It is a natural extension of the predict-correct technique, then, to modify the integration process slightly as we proceed. When we predict with the equation

$$p_{n+1} = y_{n-1} + 2\Delta x y_n \tag{6-51}$$

we might immediately modify the value of this prediction, using the previous value of the predict-correct difference and the formula

$$m_{n+1} = p_{n+1} - \tfrac{4}{5}(p_n - c_n) \tag{6-52}$$

Then we use the differential equation to compute the modified derivative:

$$m'_{n+1} = f(x_{n+1}, m_{n+1}) \tag{6-53}$$

which is then corrected by way of

$$c_{n+1} = y_n + \frac{\Delta x}{2}(m'_{n+1} + y'_n) \tag{6-54}$$

leading to the final value of y_{n+1}:

$$y_{n+1} = c_{n+1} + \tfrac{1}{5}(p_{n+1} - c_{n+1}) \tag{6-55}$$

Clearly, this procedure of predicting, modifying, correcting, and modifying again is about the extent to which we can go in solving differential equations on the pocket calculator. More advanced methods become too cumbersome.

6-9 OTHER USEFUL INDEFINITE NUMERICAL INTEGRATION FORMULAS

A number of commonly used predict-correct algorithms are convenient for pocket calculator solution of ordinary differential equations. The procedure, of course, is always the same. A data table for numerically evaluating the solution of the differential equation is prepared and then the integration formulas are used directly as written. Writing them in alternate forms does not buy much in the way of reduced number of key strokes or of data entries in the actual integration process.

The two most popular point slope formulas are the Euler predictor or midvalue predictor formulas:

$$y_{n+1} = y_n + \Delta x y'_n, \qquad (\epsilon \sim \Delta x^2)$$

$$y_{n+1} = y_{n-1} + 2\Delta x y'_n, \qquad (\epsilon \sim \Delta x^3)$$

They are usually used in conjunction with the trapezoidal corrector formula:

$$y_{n+1} = y_n + \frac{\Delta x}{2}(y'_{n+1} + y'_n), \qquad (\epsilon \sim \Delta x^3)$$

Another popular and extensively used predict-correct method is the

Adams method. Adams' predictor and corrector formulas are given by

$$y_{n+1} = y_n + \frac{\Delta x}{24}(55y'_n - 59y'_{n-1} + 37y'_{n-2} - 9y'_{n-3}), \qquad (\epsilon \sim \Delta x^5)$$

$$y_{n+1} = y_n + \frac{\Delta x}{24}(9y'_{n+1} + 19y'_n - 5y'_{n-1} + y'_{n-2}), \qquad (\epsilon \sim \Delta x^5)$$

These four-point formulas obviously require a substantial number of operations on the pocket calculator if done manually. In fact, each step involves at least 22 key strokes not including the derivative evaluation (which is problem dependent). The author has integrated a number of differential equations using Adams' formulas, but they have all been first-order differential equations (though of a complex nonlinear nature); their evaluation (though time consuming) can be done conveniently because of the fairly large step size that can be taken for equivalent accuracy with the point slope formulas. The numerical stability of these methods and the roundoff error associated with the alternating sign of the coefficients lead to difficulties, however; hence the lower-order integrators are recommended for manual numerical integration on the pocket calculator. The programmable calculator, on the other hand, can conveniently use the higher-order integration formulas and take advantage of their higher-order accuracy. The higher-order functions are therefore discussed here.

Runge-Kutta Methods

The Runge-Kutta methods are based on implicitly developing increasingly higher orders of Taylor series expansions of a function through combinations of the derivatives of a function numerically evaluated on certain intervals of the independent variable. The Runge-Kutta methods are yet another variant using the Taylor series expansion method and thus are limited in the sense that, if the integrand is not Taylor series expandable or is to be evaluated across a discontinuity, the location of the discontinuity must be determined and the solution is computed up to the discontinuity and then restarted at the discontinuity. The advantage of the Runge-Kutta methods is that they require no starting values.

The second-order Runge-Kutta method is given by

$$y_{n+1} = y_n + \tfrac{1}{2}(k_1 + k_2), \qquad (\epsilon \sim \Delta x^3) \tag{6-56}$$

where,

$$k_1 = \Delta x f(x_n, y_n)$$

$$k_2 = \Delta x f(x_n + \Delta x, y_n + k_1)$$

The Runge-Kutta methods use Euler integration at each step. Thus, to evaluate equation 6-56, it is necessary to compute both k_1 and k_2. To compute k_2, the predicted value of y $(y_n + k_1)$ must be evaluated. It is apparent that this is equivalent to Euler's method. Thus the procedure consists of first using Euler's method to compute the first estimate of y_{n+1}, which is then used along with $x_n + \Delta x$ to compute the value of the derivative at $x(n + \Delta x)$ to get k_2. Then equation 6-56 is formed, using k_1 and k_2.

Another form of Runge-Kutta's second-order equation is

$$y_{n+1} = y_n + k_2, \qquad (\epsilon \sim \Delta x^3)$$

$$k_1 = \Delta x f(x_n, y_n), \qquad\qquad\qquad (6\text{-}57)$$

$$k_2 = \Delta x f\left(x_n + \frac{\Delta x}{2}, y_n + \frac{k_1}{2}\right)$$

In this form of the Runge-Kutta equation, k_1 is employed to make a half step from x_n to $x_n + \frac{\Delta x}{2}$, where y_n is evaluated as $y_n + \frac{k_1}{2}$. Then the derivative at this midvalue, defined by $x_n + \frac{\Delta x}{2}$ is computed and used to estimate the midvalue rate from which k_2 is calculated. Then equation 6-57 is numerically evaluated using only k_2. Again, first Euler integration must be used to make the first half step, and the first full step is taken by means of the midvalue estimates of the rate on the interval.

Another Runge-Kutta method is also given in two forms. One is

$$y_{n+1} = y_n + \frac{k_1}{6} + \tfrac{2}{3} k_2, \qquad (\epsilon \sim \Delta x^4)$$

$$k_1 = \Delta x f(x_n, y_n)$$

$$k_2 = \Delta x f\left(x + \frac{\Delta x}{2}, y_n + \frac{k_1}{2}\right)$$

$$k_3 = \Delta x f(x_n + \Delta x, y_n + 2k_2 - k_1)$$

This is the most popular and convenient form of third-order Runge-Kutta integration, and is used by Hewlett-Packard in its Math Pack 1-36A solution to the first-order differential equation for its programmable pocket calculator.

Another form of third-order Runge-Kutta is

$$y_{n+1} = y_n + \frac{k_1}{4} + \frac{3}{4}k_3, \qquad (\epsilon \sim \Delta x^4)$$

$$k_1 = \Delta x f(x_n, y_n)$$

$$k_2 = \Delta x f\left(x_n + \frac{\Delta x}{3}, y_n + \frac{k_1}{3}\right)$$

$$k_3 = \Delta x f\left(x_n + \frac{2\Delta x}{3}, y_n + \frac{2k_2}{3}\right)$$

Though these equations look formidable, their solution involves only three steps of Euler integration at the most. Again, the advantage is that they require no starting values.

The two most popular forms of the Runge-Kutta fourth-order numerical integration are the following:

$$y_{n+1} = y_n + \frac{k_1}{6} + \frac{k_2}{3} + \frac{k_3}{3} + \frac{k_4}{6}, \qquad (\epsilon \sim \Delta x^5)$$

$$k_1 = \Delta x f(x_n, y_n)$$

$$k_2 = \Delta x f\left(x_n + \frac{\Delta x}{2}, y_n + \frac{k_1}{2}\right)$$

$$k_3 = \Delta x f\left(x_n + \frac{\Delta x}{2}, y_n + \frac{n_2}{2}\right)$$

$$k_4 = \Delta x f(x_n + \Delta x, y_n + k_3)$$

and

$$y_{n+1} = y_n + \frac{k_1}{8} + \frac{3k_2}{8} + \frac{3k_3}{8} + \frac{k_4}{8}, \qquad (\epsilon \sim \Delta x^5)$$

$$k_1 = \Delta x f(x_n, y_n)$$

$$k_2 = \Delta x f\left(x_n + \frac{\Delta x}{3}, y_n + \frac{k_1}{3}\right)$$

$$k_3 = \Delta x f\left(x_n + \frac{2\Delta x}{3}, y_n + k_2 - \frac{k_1}{3}\right)$$

$$k_4 = \Delta x f(x_n + \Delta x, y_n + k_3 - k_2 + k_1)$$

In all of the methods presented in this section the differential equation is assumed to be of the first order and generally written in the form $y' = f(x,y,)$. Since an nth-order differential equation can be written in terms of n first-order differential equations, these methods are applicable to systems of equations or to higher-order equations.

A number of specific methods are available for higher-order differential equations, and in those cases special predict-correct algorithms can be developed. Although, for general-purpose computing, they are not very useful for numerical evaluation of the solution of the differential equation, they simplify the number of calculations on the pocket calculator. For example, Milne's predictor-corrector algorithms for first-order differential equations take the forms

$$\left.\begin{matrix} P \\ C \end{matrix}\right\}\left\{\begin{matrix} y_{n+1} = y_{n-3} + \dfrac{4\Delta x}{3}(2y'_n - y'_{n-1} + 2y'_{n-2}) \\ y_{n+1} = y_{n-1} + \dfrac{\Delta x}{3}(y'_{n-1} + 4y'_n + y'_{n+1}) \end{matrix}\right. \qquad (\epsilon \sim \Delta x^5)$$

$$\left.\begin{matrix} P \\ C \end{matrix}\right\}\left\{\begin{matrix} y_{n+1} = y_{n-5} + \dfrac{3\Delta x}{10}(11y'_n - 14y'_{n-1} + 26y'_{n-2} - 14y'_{n-3} + 11y'_{n-4}) \\ y_{n+1} = y_{n-3} + \dfrac{2\Delta x}{45}(7y'_{n+1} + 23y'_n + 12y'_{n-1} + 32y'_{n-2} + 7y'_{n-3}) \end{matrix}\right.$$

$$(\epsilon \sim \Delta x^7)$$

The equivalent accuracy Milne predictor-corrector formulas, for second-, and third-order differential equations, are written as follows:

$$\left.\begin{matrix} P \\ C \end{matrix}\right\}\left\{\begin{matrix} y_{n+1} = y_{n-2} + 3(y_n - y_{n-1}) + \Delta x^2(y''_n - y''_{n-1}), \qquad (\epsilon \sim \Delta x^5) \\ y_{n+1} = y_n + \dfrac{\Delta x}{2}(y'_{n+1} + y'_n) - \dfrac{\Delta x^2}{12}(y''_{n+1} - y''_n) \end{matrix}\right.$$

$$\left.\begin{matrix} P \\ C \end{matrix}\right\}\left\{\begin{matrix} y_{n+1} = y_{n-2} + 3(y_n - y_{n-1}) + \dfrac{\Delta x^3}{2}(y'''_n - y'''_{n-1}) \\ y_{n+1} = y_n + \dfrac{\Delta x}{2}(y'_{n+1} + y'_n) - \dfrac{\Delta x^2}{10}(y''_{n+1} - y''_n) \end{matrix}\right. \qquad (\epsilon \sim \Delta x^7)$$

$$+ \frac{\Delta x^3}{120}(y'''_{n+1} + y'''_n)$$

For systems of differential equations of the form

$$y' = f(x,y,z), \quad z' = g(x,y,z)$$

second-order Runge-Kutta can be written as

$$\begin{cases} y_{n+1} = y_n + \dfrac{k_1}{2} + \dfrac{k_2}{2} \\[2mm] z_{n+1} = z_n + \dfrac{l_1}{2} + \dfrac{l_2}{2} \end{cases} \qquad (\epsilon \sim \Delta x^3)$$

$$k_1 = \Delta x f(x_n, y_n, z_n)$$

$$l_1 = \Delta x g(x_n, y_n, z_n)$$

$$k_2 = \Delta x f(x_n + \Delta x, y_n + k_1, z_n + l_1)$$

$$l_2 = \Delta x g(x_n + \Delta x, y_n + k_1, z_n + l_1)$$

Fourth-order Runge-Kutta for this system of equations takes the form

$$y_{n+1} = y_n + \frac{k_1 + 2k_2 + 2k_3 + k_4}{6}$$

$$z(n+1) = z_n + \frac{l_1 + 2l_2 + 2l_3 + l_4}{6}$$

$$k_1 = \Delta x f(x_n, y_n, z_n)$$

$$l_1 = \Delta x g(x_n, y_n, z_n)$$

$$k_2 = \Delta x f\left(x_n + \frac{\Delta x}{2}, yn + \frac{k_1}{2}, z_n + \frac{l_1}{2}\right)$$

$$l_2 = \Delta x g\left(x_n + \frac{\Delta x}{2}, y_n + \frac{k_1}{2}, z_n + \frac{l_1}{2}\right)$$

$$k_3 = \Delta x f\left(x_n + \frac{\Delta x}{2}, y_n + \frac{k_2}{2}, z_n + \frac{l_2}{2}\right)$$

$$l_3 = \Delta x f\left(x_n + \frac{\Delta x}{2}, y_n + \frac{k_2}{2}, z_n + \frac{l_2}{2}\right)$$

$$k_4 = \Delta x f(x_n + \Delta x, y_n + k_3, z_n + l_3)$$

$$l_4 = \Delta x f(x_n + \Delta x, y_n + k_3, z_n + l_3)$$

Another special form of second-order differential equation is

$$y'' = f(x, y, y')$$

Milne's predictor corrector method for these types of second-order equations is as follows:

$$y'_{n+1} = y'_{n-3} + \frac{4\Delta x}{3}(2y''_{n-2} - y''_{n-1} + 2y''_n)$$

$$y'_{n+1} = y'_{n-1} + \frac{\Delta x}{3}(y''_{n-1} + 4y''_n + y''_{n+1}) \qquad (\epsilon \approx \Delta x^5)$$

The single-step self-starting Runge-Kutta method takes the form

$$y_{n+1} = y_n + \Delta x y'_n + \frac{\Delta x}{6}(k_1 + k_2 + k_3)$$

$$y'_{n+1} = y'_n = \frac{1}{6}(k_1 + 2k_2 + 2k_3 + k_4) \qquad (\epsilon \sim \Delta x^5)$$

$$k_1 = \Delta x f(x_n, y_n, y'_n)$$

$$k_2 = \Delta x f\left(x_n + \frac{\Delta x}{2}, y_n + \frac{\Delta x}{2}y'_n + \frac{\Delta x k_1}{8}, y'_n + \frac{k_1}{2}\right)$$

$$k_3 = \Delta x f\left(x_n + \frac{\Delta x}{2}, y_n + \frac{\Delta x y'_n}{2} + \frac{\Delta k_1}{8}, y'_n + \frac{k_2}{2}\right)$$

$$k_4 = \Delta x f\left(x_n + \Delta x, y_n + \Delta x y'_n + \frac{\Delta x k_3}{2}, y'_n + k_3\right)$$

For second-order differential equations,

$$y'' = f(x, y)$$

Milne's method takes the forms

$$y_{n+1} = y_n + y_{n-2} - y_{n-3} + \frac{\Delta x^2}{4}(5y''_n + 2y''_{n-1} + 5y''_{n-2})$$

$$y_{n+1} = 2y_n - y_{n-2} + \frac{\Delta x^2}{12}(y''_{n+1} + 10y''_n + y''_{n-1}) \qquad (\epsilon \sim \Delta x^6)$$

The Runge-Kutta method appears as

$$y_{n+1} = y_n + \Delta x \left[y'_n + \left(\frac{k_1 + 2k_2}{6} \right) \right]$$

$$y'_{n+1} = y'_n + \frac{k_1}{6} + \frac{2k_2}{3} + \frac{k_3}{6}, \qquad (\epsilon \sim \Delta x^4)$$

$$k_1 = \Delta x f(x_n y_n)$$

$$k_2 = \Delta x f \left(x_n + \frac{\Delta x}{2}, y_n + \frac{\Delta x}{2} y'_n + \frac{\Delta x}{8} k_1 \right)$$

$$k_3 = \Delta x f \left(x_n + \Delta x, y_n + \Delta x y'_n + \frac{\Delta x}{2} k_2 \right)$$

In the second-order forms, the Runge-Kutta algorithms involve the numerical evaluation of rates by Euler integration of the second-order differential equation. For the programmable pocket calculator with limited memory, these alternate forms of numerically evaluating indefinite integrals are particularly useful because they dispense with computing the two first-order differential equations that would be required to make up the second-order equation in the more general first-order indefinite integration formulas.

6-10 T-INTEGRATION

T-Integration (tunable integration) is a new flexible integration concept that permits the integration formula to be tuned to the system of equations it is solving. In its simplest form it is written:

$$y_n = y_{n-1} + \lambda T [\gamma \dot{y}_n + (1 - \gamma) \dot{y}_{n-1}]$$

This equation is based on adjusting the phasing of the integration so as to satisfy the mean value theorem (as opposed to numerical integration algorithms based on analytical substitution techniques). The parameter γ controls the amount of transport lead (or lag) imposed on the integrand of the integral. For example, $\gamma = -\frac{1}{2}$ means that the integrand has been time delayed one sample period, while $\gamma = +\frac{3}{2}$ implies that the integrand is time advanced one sample period. Since in the numerical integration of differential equations the solution point is not known before it is computed, and thus cannot be made part of the integral, it is estimated using an extrapolation formula. It is apparent from the equation for this integrator that the weight of the two coefficients in the integrand is performing the interpolation/extrapolation operation. Therefore, an approximate equivalent form of the T-integration equation is

$$y_n = y_{n-1} + \lambda T[(\gamma + 1)\dot{y}_{n-1} - \gamma \dot{y}_{n-2}]$$

In its application, T-integration is usually used in the following manner:

1. The differential equation and the interval over which its solution is to be computed are defined together with its initial condition.

2. The integration step size is set equal to one-tenth of the interval size or one-tenth of the shortest period in the oscillations of the solution expected for the differential equation, whichever is smaller. If the solution of the differential equation is expected to be exponential, or smooth, or of monotonic nature, the step size is set at one-tenth the interval over which the solution is to be evaluated.

3. If the integration is an open-loop process, that is, the integrand is not a function of the integral, then γ is set equal to $\frac{1}{2}$ and the differential equation is numerically integrated. If, however, the integrand is a function of the integral, γ is set equal to $\frac{3}{5}$. When closed-loop integration is performed, the sequence of solutions can be plotted and the points connected with straight lines.

4. The solution can then be compared with check cases that may be run at smaller integration step sizes or with empirically prepared check cases. It is generally found that the solutions generated from the use of the T-integrator "lead" the check case by approximately one integrating interval or slightly less.

Note, however, that the *dynamics* of the solution prepared with the T-integrator match the *dynamics* of any check case. That is, although the T-integrator, which is a low-order integrator, permits accurate simulation of the *dynamics* of a discrete process, it does so at the sacrifice of a slight phase error. Nevertheless, in many engineering applications it is sufficient for determining, for example, peak overshoot, natural frequency, damping, resonant frequencies, and conditions of dynamic instability, which are the purpose of the analysis. In general, it must be remembered that all of the integration formulas presented here are usually not for generating of numbers to six places, but rather for solving problems and understanding the dynamics of processes for the purposes of design, test, and evaluation, or all of them.

6-11 REFERENCES

For this chapter use Richard Hamming's *Numerical Methods for Scientists and Engineers* (McGraw-Hill, 1973), Chapters 21 through 24, and J. M. Smith's, "Recent Developments in Numerical Integration," *ASME Journal of Dynamic Systems, Measurement, and Control*, March 1974, pages 61-70.

CHAPTER 7

LINEAR SYSTEMS
SIMULATION

7-1 INTRODUCTION

The analysis of linear constant coefficient systems is important because
they are frequently encountered in the design of continuous processes. The
dynamic characteristics of a linear systems response to known types of
forcing functions are usually studied when setting the parameters for a
system design. In this chapter we discuss the synthesis of recursion formulas by which the response of a linear dynamic process to sampled values of
its forcing function can be conveniently computed. We tailor numerical
integration and other discrete approximation methods for computing the
dynamics of continuous processes to pocket calculator analysis. On a
pocket calculator it is much easier to iterate a recursion formula to
compute the dynamics of a process than it is to actually conduct the
numerical integration of the process. Under some circumstances (when
there are no hard nonlinearities, such as limits, hysteresis, and dead zones),
it is quite easy to develop the recursion formulas from the integration
formulas, thus eliminating many steps in the computing of the solution to
high-order differential equations. In fact, the number of key strokes can be
reduced 80% with recursion formulas (difference equations) as compared
to that needed in direct numerical integration of a differential equation.

7-2 DERIVATION OF DIFFERENCE EQUATIONS BY NUMERICAL INTEGRATION SUBSTITUTION

Examples of many numerical integration formulas have already been
discussed, such as Euler's integration formula, rectangular integration,

180

trapezoidal integration, T-integration, and a number of predict-correct formulas. We used the differential equation to numerically evaluate the derivatives at the initial condition and then from the starting values in the integration formulas we predicted the solution to the differential equation in the neighborhood of the initial conditions.

Another use of a numerical integration formula is to form a difference equation. Consider the first-order constant coefficient differential equation

$$\tau \dot{x} + x = Q \tag{7-1}$$

where

$$x = x(t)$$

$$Q = Q(t)$$

$$\tau = \text{a constant}$$

Now consider the Euler integration formula

$$x_n = x_{n-1} + T\dot{x}_{n-1} \tag{7-2}$$

We can solve for the rate in the integration formula, using the differential equation, as follows:

$$\dot{x}_{n-1} = \frac{1}{\tau}(Q_{n-1} - x_{n-1}) \tag{7-3}$$

This can be substituted back into the numerical integration formula

$$x_n = x_{n-1} + \frac{T}{\tau}(Q_{n-1} - x_{n-1}) \tag{7-4}$$

which, when simplified, gives the difference equation

$$x_n = \left(1 - \frac{T}{\tau}\right)x_{n-1} + \frac{T}{\tau}Q_{n-1} \tag{7-5}$$

This recursion formula computes, for example, the 100th step in the solution of the differential equation on the basis of data generation on the 99th step. The indices in the recursion formula keep track of the iteration that is being computed when solving the differential equation. They also indicate the approximate time at which the solution value will compare with $x(t)$, that is, $t = nT$ if the solution begins at $T \cong 0$. We shall see later that $t \neq nT$, but it is sufficiently close to approximately label the time in the sequence of solution values of the difference equation.

The use of recursion formulas in solving differential equations has two advantages. They reduce the number of key strokes needed to evaluate the solution of the differential equation on the pocket calculator. And in linear constant coefficient processes they permit the use of *implicit* integration formulas. It is these formulas in which the rates of a state variable are a function of the state itself. The trapezoidal integration formula is an example:

$$x_{n+1} = x_n + \frac{T}{2}(\dot{x}_{n+1} + \dot{x}_n) \tag{7-6}$$

Trapezoidal integration computes the $n+1$ value of x based on the $n+1$ value of $\dot{x}$. However, evaluating $\dot{x}$ in the differential equation requires x_{n+1}. This results in an implicit equation, whose solution is a function of itself. When implicit integration formulas are used to derive difference equations, the implicit equation can be solved *algebraically*. For example, consider the implicit Euler integration (rectangular integration), which takes the form

$$x_n = x_{n-1} + T\dot{x}_n \tag{7-7}$$

By way of our first-order differential equation, we obtain

$$\dot{x}_n = \frac{1}{\tau}(Q_n - x_n) \tag{7-8}$$

which, when substituted back into the implicit rectangular integration formula, gives the difference equation

$$x_n = x_{n-1} + \frac{T}{\tau}(Q_n - x_n) \tag{7-9}$$

Note that this equation is still in implicit form; that is, x_n is a function of itself. However, it can be solved algebraically as follows:

$$x_n + \frac{T}{\tau}x_n = x_{n-1} + \frac{T}{\tau}Q_n$$

$$\left(1 + \frac{T}{\tau}\right)x_n = x_{n-1} + \frac{T}{\tau}Q_n$$

$$\therefore x_n = \left(\frac{1}{1 + T/\tau}\right)x_{n-1} + \left(\frac{T/\tau}{1 + T/\tau}\right)Q_n \tag{7-10}$$

Let us now compare the implicit and explicit Euler difference equations from the standpoints of numerical stability, numerical error, the manner in which the differential equation seeks its final value, and their implementation on the pocket calculator.

The stability of these first-order difference equations is completely determined by the magnitude of the first coefficient in the difference equation. That is, if the term

$$\frac{1}{1 + T/\tau} \qquad\qquad 1 + \frac{T}{\tau}$$

$$\text{Implicit Integration} \qquad \text{Explicit integration}$$

exceeds ± 1, the difference equation becomes unstable. For example, if $a = 2$ in the difference equation $y_n = ay_{n-1}$, the difference equation takes on the solution values shown in Table 7-1. Note, however, that at $a = 0.9$ the difference equation is stable, as shown in Table 7-2. The stability criterion in first-order difference equations generally is that the magnitude of a be less than or equal to 1. Now, notice the first-order difference equation that is generated with the explicit Euler integration.

Our aim here is to determine the conditions under which the integration step size and the system's time constant allows a stable difference equation, rather than leading to numerical instability. We, therefore, first determine the conditions under which the magnitude of a is less than or equal to 1. That is

$$\left| 1 - \frac{T}{\tau} \right| \leqslant 1$$

Solving the inequality for T/τ, we see that the region of stability for the

Table 7-1 Unstable Response of the Difference Equation $y_n = ay_{n-1}$ where $a = 2$

n	y_n
1	1
2	2
3	4
4	8
5	16
.	.
.	.
.	.

**Table 7-2 Stable Response of the Difference
Equation $y_n = a Y_{n-1}$ where $a = 0.9$**

n	y_n
1	1
2	0.9
3	0.81
4	0.729
5	0.6561
$\vdots$	$\vdots$

difference equation derived with explicit Euler integration is

$$0 \leqslant \frac{T}{\tau} \leqslant 2$$

On examining the difference equation derived with rectangular integration (implicit Euler integration), we see that the condition under which

$$\left| \frac{1}{1 + T/\tau} \right| \leqslant 1$$

is

$$0 \leqslant \frac{T}{\tau}$$

Clearly, the difference equation developed with rectangular integration is much more stable than that generated by Euler explicit integration. This is a specific example of the more general result that implicit integration of constant coefficient linear differential equations leads to intrinsically more stable difference equations than do those developed with explicit integration formulas. We therefore concentrate on the use of implicit integration formulas in developing difference equations for simulating continuous processes.

Now, let us look at the accuracy of these simulating difference equations. Table 7-3 shows the sequence of solution values for the explicit and implicit difference equation's response to a unit step. The greatest precision is clearly achieved with the implicit formula. These difference equations were tested for an integration step size divided by the time constant equal

Table 7-3 Comparison of Implicit and Explicit Integration-Derived Difference Equations when $T/\tau = 1.5$

Normalized Time	Exact $x(nT)$	Implicit $x(nT)$	Error	Explicit $x(nT)$	Error
$\dfrac{T}{\tau} = 0$	0	0	0	0	0
$\dfrac{T}{\tau} = 1.5$	0.776	0.600	−0.176	1.50	+0.90
$\dfrac{T}{\tau} = 3.0$	0.950	0.840	−0.110	0.75	−0.20
$\dfrac{T}{\tau} = 4.5$	0.990	0.936	−0.054	1.125	+0.135

to $\frac{3}{2}$, which challenges the stability of the Euler-derived difference equation. Both equations appear to be stable. However, the implicit difference equation is obviously more accurate than is the explicit equation. This is another special case of a general property of difference equations derived with implicit integration to simulate linear constant coefficient systems. The implicitly derived difference equations are generally more accurate than those derived explicitly.

Finally, let us examine the steady state that all these difference equations achieve. To do so, we must examine the nonhomogeneous equation (since in a homogeneous equation all the end conditions of the steady states approach zero, thus making comparison impossible). For the continuous and discrete equations, the step response has the forms shown below:

$$y = Q(1 - e^{-t/\tau}), \quad y_n = \left(1 - \frac{T}{\tau}\right)y_{n-1} + \frac{T}{\tau}Q_{n-1}$$
$$\text{Exact} \qquad\qquad\qquad \text{Explicit}$$

$$y_n = \left(\frac{1}{1 + T/\tau}\right)y_{n-1} + \left(\frac{T/\tau}{1 + T/\tau}\right)Q_n$$
$$\text{Implicit}$$

In the steady state

$$x_n = x_{n-1}$$

Thus we can write the final value as follows:

$$\lim_{t \to \infty} y(t) = Q \quad \underset{\text{Explicit}}{y_n = Q_n = Q_{n-1}} \quad \underset{\text{Implicit}}{y_n = Q_n}$$
$$\underset{\text{Exact}}{}$$

In summary: both the explicitly and implicitly derived difference equations achieve the same final value for the unit step forcing functions which is the final value for the true continuous process. But an implicitly derived recursion formula is more stable and more accurate than its explicitly derived counterpart.

Now let us compare the numerical integration of the differential equation with that achieved by using the recursion formula. The sequence of key strokes required to perform the numerical integration of this first-order differential equation

$$\tau \dot{x} + x = Q$$

using Euler's integration formula is shown in Table 7-4. Table 7-5 shows the key strokes involved in the use of the difference equation.

Table 7-4 Typical Key Stroke Sequences for Numerically Integrating $\tau \dot{x} + x = Q$

Reverse-Polish	Algebraic
(Q_{n-1})	(Q_{n-1})
$RCL\,1 \quad \leftarrow x(0)$ prestored	$-$
$-$	$RCL \quad \leftarrow x(0)$ prestored
$RCL\,2 \quad \leftarrow \tau$ prestored	$\times$
	$\left(\dfrac{1}{\tau}\right)$
$\div$	$\times$
$RCL\,3 \quad \leftarrow T$ prestored	(T)
$\times$	$+$
$RCL\,1$	RCL
$+$	$=$
$\boxed{x_n}$	$\boxed{x_n}$
STO 1	STO

( )→data entry.

▢ →output.

Table 7-5 Typical Key Stroke Sequence for Difference Equation Evaluation of $\tau\dot{x} + x = Q$

Reverse-Polish	Algebraic
(Q_{n-1})	(Q_{n-1})
$RCL\,1 \;\leftarrow\!\left(\dfrac{T}{\tau}\right)$ prestored	$\times$
$\times$	$\left(\dfrac{T}{\tau}\right)$
$RCL\,2 \;\leftarrow x(0)$ prestored	$+$
$RCL\,3 \;\leftarrow(1-\dfrac{T}{\tau})$ prestored	$RCL \;\leftarrow x(0)$ prestored
$\times$	$\times$
$+$	$\left(1\text{-}\dfrac{T}{\tau}\right)$
$\boxed{x_n}$	$=$
$STO\,2$	$\boxed{x_n}$
	STO

$(\;\;)\rightarrow$data entry.

$\boxed{}\rightarrow$output.

And Table 7-6 summarizes the key strokes involved in the precalculation and iteration through the first step, the first 10 steps, and then 20 steps.

We see that even for these simple integrators in this simple differential equation the reduction in key strokes using the recursion formulas (8.4% and 4.2%) is important enough to warrant the use of recursion formulas. A greater number of keystrokes is saved when recursion formulas are used to simulate high-order linear systems.

These recursion formulas are particularly useful in evaluating a system's response to an arbitrary forcing function. Provided that the integration step size is small compared with the largest period of interest in the oscillations of the forcing function, the recursion formulas can efficiently evaluate the system's response to an arbitrary forcing function on the pocket calculator and in particular on the programmable calculator, where the implicit difference equations take up much less memory than do the numerical integration formulas and direct numerical integration.

A possible difficulty associated with the implicit integration formula for evaluating the response of a system to an arbitrary forcing function is its assumption that the forcing function is known at time nT, in order to

Table 7-6 Number of Key Strokes (Worst-Case)*^{a} Required to Simulate the Continuous Process $\dot{y} + ky = f$

	Numerical Integration		Euler-Derived Recursion Formula	
	Reverse-Polish	Algebraic	Reverse-Polish	Algebraic
Number of precalculation key strokes	39	13	39	13
Number of key strokes for the first iteration	22	47	20	45
Subtotal	(61)	(60)	(59)	(58)
Number of key strokes for the tenth iteration	220	470	200	450
Subtotal	(281)	(530)	(259)	(508)
Total number of key strokes for the twentieth iteration	501	1000	459	958

^{a}Assumes 13-digit Data Entries.

compute the response of the system at time n. If the forcing function is of the form

$$f = f(y, t)$$

the evaluation of f_n requires

$$f_n = f(y_n, t_n)$$

but since the difference equation is still to compute y_n, it is not yet in our table of solution values; instead we have only a tabulated value for y_{n-1}. In this case we can use an extrapolation formula to estimate y_n by way of the two past values, or we can use y_{n-1} merely as an approximation of y_n. This can be done when the forcing function's components are (from a Fourier analysis viewpoint) of lower frequency than is the natural frequency of the system described by the differential equation. To achieve this, we calculate a few values of the difference equation, assuming in evaluating the forcing function that $y_n \approx y_{n-1}$ and generating the first few terms of the forcing function f, and use a difference table to evaluate whether f is changing rapidly. If the change is rapid, we simply use an interpolation formula to make a first estimate of y_n based on y_{n-1} and y_{n-2}. The author rarely finds it necessary to use the extrapolation scheme in the practical evaluation of the solution to differential equations.

This technique of deriving difference equations to simulate continuous dynamic processes is extremely useful for simulating the dynamics of nonlinear processes. One problem is that most implicit difference equations cannot be solved for nonlinear differential equations. That is, the implicit equation is a nonlinear equation, and usually only iterative techniques can be used to solve it. However, the explicit difference equation is easily derived and easily put in a form that can be quickly evaluated on the pocket calculator, as opposed to numerically integrating the nonlinear equation.

7-3 STABLE DIFFERENCE EQUATIONS

Recursion formulas for simulating continuous dynamic processes can also be derived by assuming a difference equation of the same order as the differential equation to be simulated. Then match the roots of the difference equation with the roots of the differential equation and include an "adjustment factor" so as to match the final value of the difference equation with the final value of the differential equation. All that remains, then, is to add another "adjustment factor" to match the phasing of the difference equation to the phasing of the solution to the differential equation. For example, consider again the simple first-order constant coefficient continuous process

$$\tau \dot{x} + x = Q$$

Assume that this equation has a solution of the homogeneous equation

$$x = e^{ts} \tag{7-11}$$

On substitution, we can derive the indicial equation as

$$(\tau s + 1)e^{ts} = 0 \tag{7-12}$$

which has the characteristic root

$$s = -\frac{1}{\tau} \tag{7-13}$$

Clearly, then, the solution to the homogeneous differential equation takes the form

$$x = c_1 e^{-t/\tau} \tag{7-14}$$

The solution to the nonhomogeneous equation can be derived with the convolution integral where the solution of the homogeneous equation is convolved with the forcing function:

$$x = \int_0^t Q(k) e^{[(k-t)/\tau]} dk \tag{7-15}$$

The complete solution to the differential equation then takes the form

$$x = e^{-t/\tau} \left\{ \int_0^t Q(k) e^{k/\tau} dk + c_1 \right\} \tag{7-16}$$

Similar procedures can be followed for higher-order differential equations, using either time-domain analysis, Laplace transform theory, or even Z-transform theory.

Let us assume that we are going to simulate this continuous process with a difference equation whose roots and final value match those of the continuous process. We assume a difference equation:

$$x_n = a x_{n-1} + b Q_n \tag{7-17}$$

A solution to the homogeneous difference equation is of the form

$$x_n = c_1 e^{-nT/\tau} \tag{7-18}$$

Upon substitution, it leads to the indicial equation for the difference equation:

$$c_1 e^{-nT/\tau} (1 - a e^{T/\tau}) = 0 \tag{7-19}$$

Thus for the roots of the difference equation to match the roots of the differential equation, we require that

$$a = e^{-T/\tau} \tag{7-20}$$

This determines the coefficient in the difference equation that accomplishes the pole matching between the difference and differential equations. It is clear that the solution to the homogeneous difference equation is

$$x_n = e^{-T/\tau} x_{n-1} \tag{7-21}$$

This procedure has now guaranteed that the dynamics of the difference equation will match the dynamics of the differential equation because their roots are equivalent and they will generate equivalent solution values as seen by the exponential decay of both. What remains is to compute the final value of the difference equation and match it with that of the differential equation. The procedure here is more straightforward in that the nonhomogeneous difference equation takes the form

$$x_n = e^{-T/\tau} x_{n-1} + bQ_n \tag{7-22}$$

where the steady state of the root-matched difference equation is achieved when

$$Q_n = Q_{n-1}$$

$$x_n = x_{n-1}$$

Then

$$x_n = \frac{b}{1 - e^{-T/\tau}} Q_n \tag{7-23}$$

By including the final value adjustment factor

$$b = 1 - e^{-T/\tau} \tag{7-24}$$

we can make the difference equation achieve the same final value as the differential equation. Thus the simulating difference equation takes the form

$$x_n = e^{-T/\tau} x_{n-1} + (1 - e^{-T/\tau}) Q_n \tag{7-25}$$

Notice that the homogeneous solution of this difference equation matches the homogeneous solution of the differential equation *exactly*. Also, it generates a sequence of solutions that are exact for the step response $(Q(t) = U(t))$ and will generate solutions that are a good approximation of the differential equation's response to an arbitrary forcing function. Also notice that this difference equation is incapable of going unstable, regardless of the integration step size (because the term $e^{-T/\tau}$ is always less than 1 no matter how big T gets provided that $\tau > 0$).

From the tabulated values it may appear that there is significant error in the solutions generated with the dynamics-matched difference equation and that generated with the continuous differential equation. However, equation 7-25 makes it clear that the difference equation solutions are lagging the continuous solutions. The dynamics are usually identical to the differential equation except for this effect of phase shift. Of course, we could reduce the step size to bring the two curves closer, but, this is not an efficient or correct approach to reducing this kind of error. Or we can compensate for this phasing error (transport delay) by determining with interpolation at what time the sequence of solutions generated by the difference equation matches the differential equation and then including that transport time in the tabulation of the sequence of solutions generated in the difference equation. Suppose that we know that for the fourth entry in a table of solution values the true continuous solution lies somewhere between $t = 3T$ and $4T$. Using inverse interpolation, we can determine the time at which the discrete solution matches the continuous solution and then arbitrarily select that time as the reference time from which we count nT intervals.

It is important to remember that the solution values generated with difference equations and even with numerical integration formulas are operating at a problem time which is different from the sequence of times nT. That is, problem time in a discrete approximation of a continuous time process is different from the sequence of values nT. Hence the indices in the recursion formulas represent the number of iterations, not time nT. The analyst must determine the actual timing of the sequence of solution values in order to compare them with a true continuous-time check case. It is the author's experience that many engineers and programmers, on large digital computers as well as on pocket calculators, overlook this problem of timing and try to compare continuous and discrete computing processes at times nT instead of recognizing that numerical integration is *an approximating process*. There is a timing problem also in the synthesis of simulating difference equations by dynamics matching. In fact, discrete systems are different in their operation on the flow of information in feedback loops, whether in numerical integrators or in difference equations. Thus the phasing of the two sequences of values between continuous and discrete dynamic processes must be taken into account by the analyst. The problem really arises only with large integration step sizes, but it is precisely then that efficiency is at a premium and, especially on the pocket calculator workload is substantially reduced from that for an integration step size only half as long.

Once again we find that pocket calculator may be the analytical tool for teaching the difference between discrete and continuous systems dynamics and the simulation of one with the other.

A few of the commonly encountered linear processes and their simulating difference equations using dynamics matching methods are tabulated in Table 7-7.

It is *imperative* that when the simulating difference equations are used the table of solution values be referenced to the number of iterations

Table 7-7 Difference Equations for Commonly Encountered Linear Constant Coefficient Systems

$G(s)$	$f(t)$	Difference Equations for Simulation
$\dfrac{I}{f} = \dfrac{1}{s}$	$I = \int f(t)\,dt$	$I_n = I_{n-1} + Tf_n$
$\dfrac{I}{f} = \dfrac{1}{s^2}$	$I = \int\int f(t)\,dt^2$	$I_n = 2I_{n-1} - T_{n-2} + \dfrac{T^2}{2}(f_n + f_{n-1})$
$\dfrac{y}{x} = \dfrac{1}{\tau s + 1}$	$e^{-t/\tau}$	$y_n = e^{-T/\tau}y_{n-1} + (1 - e^{-T/\tau})x_n$
$\dfrac{y}{x} = \dfrac{bs}{x+a}$	$\dfrac{-b}{a\tau}e^{-t/\tau}$	$y_n = e^{-T/\tau}y_{n-1} + be^{-}a^T(x_n - x_{n-1})$
$\dfrac{y}{x} = \dfrac{w_n^2}{s^2 + 2\zeta w_n s + w_n^2}$		$y_n = Ay_{n-1} - By_{n-2} + (1 - A + B)x_n$
		$A = 2e^{-\zeta w_n T}\cos\{w_n T(1 - \zeta^2)^{1/2}\}$
		$B = e^{-2\zeta w_n T}$
		when $0 \leqslant \zeta < 1$

through the difference equations, not to time nT. The comparison of discrete solution values with a continuous check case *involves timing considerations*, and it is the analyst's responsibility to determine the proper comparison in a manner similar to that mentioned above.

The difference equations just developed by dynamics matching methods have some very important general properties. These difference equations are intrinsically stable if the process under consideration is stable. That is, there is no sample period T to cause these equations to become unstable if the continuous process that they are simulating is stable. This is because the roots of the differential equation are matched with the roots of the difference equation; hence if the continuous process is stable, the discrete process is stable independent of sample period. Showing that the magnitude of the roots of the discrete system are less than or equal to 1 will prove this; the very way in which they are formulated shows this to be so. For example, in the first-order case that we just developed, when the roots of the discrete system are matched to the roots of the continuous system, the discrete system root is given by

$$e^{-T/\tau}$$

which will be always less than or equal to 1 for all $T > 0$ and for $\tau > 0$. The only condition on using the difference equation is that the forcing function be sampled at a rate equal to twice the highest frequencies of interest in the forcing function. More detail is given in Chapter 5 where sampling rate is discussed.

Also, the final value of the discrete difference equation will always match that of the continuous difference equation, independent of sampling period and without the final value adjustment factor. That this is so can be established by the fact that in the steady state the present and past values of the response of the system are the same. When substituted into the difference equation, the final value of the response can be computed in terms of the input forcing function, which is found to match the final value of the continuous dynamic process being simulated.

There is a limitation in the use of these simulating difference equations. Clearly, a second-order continuous system can have three different dynamic characteristics: when the two roots of the system are real and equal; when they are real and unequal; and when both are complex. The dynamics of the second-order continuous system with complex roots is damped oscillatory in nature, and the response of the system with real roots is nonoscillatory, being damped only. Each case requires different types of difference equations. It is important, then, to know where the roots are in the complex plane to determine which difference equation is to be used. This is particularly true if the coefficients in the differential equation are changing with time and are not fixed, as in the case of linear constant coefficient systems. When the coefficients are time varying, these difference equations can be used for piecewise linear constant coefficient approximation, with the results matching closely the numerically integrated solution of the time-varying differential equation. However, if the time-varying roots jump on and off the real axis, switching from one difference equation to another is necessary. That is, one difference equation simulates the dynamics of the process when the roots are real but not equal; another difference equation simulates the dynamics when the roots are real and equal; and yet another difference equation serves when the roots are complex. The choice of the appropriate set of difference equations is fairly straightforward, but note that the implicit difference equation generated in Section 7-2 does not require this changing of difference equations and thus might be more applicable from the standpoint of quickly simulating continuous processes on the pocket calculator.

7-4 VARIANCE PROPAGATION

Computing the propagation of noise through discrete linear constant coefficient processes is very easy when the noise is "almost white" and stationary. The approach is to rewrite the difference equation into a finite memory form where the response is only a function of present and past values of the forcing function. This can be easily developed in the following manner. Given

$$x_n = \sum_{i=1}^{m} a_i x_{n-i} + \sum_{j=1}^{p} b_j Q_{n-j} \tag{7-26}$$

We wish to find

$$x_n = \sum_{k=1}^{q} c_k Q_{n-k} \tag{7-27}$$

The usual approach is to rewrite $x_n = f(x_{n-1}, Q)$ in the sequence

$$x_n = f(x_{n-1}, Q)$$

$$x_n = f[f(x_{n-2}, Q)]$$

$$x_n = f[f\{f(x_{n-3}, Q)\}]$$

$$\vdots$$

Induction is then used to form the rest of the series. For example,

$$x_n = e^{-T/\tau} x_{n-1} + (1 - e^{-T/\tau}) Q_{n-1} \tag{7-28}$$

$$x_n = e^{-T/\tau} [e^{-T/\tau} x_{n-2} + (1 - e^{-T/\tau}) Q_{n-2}] + (1 - e^{-T/\tau}) Q_{n-1} \tag{7-29}$$

$$x_n = e^{-2T/\tau} x_{n-2} + (1 - e^{-T/\tau}) [Q_{n-1} + e^{-T/\tau} Q_{n-2}] \tag{7-30}$$

We can expect the next substitution to give

$$x_n = e^{-3T/\tau} x_{n-3} + (1 - e^{-T/\tau}) [Q_{n-1} + e^{-T/\tau} Q_{n-2} + e^{-2T/\tau} Q_{n-3}] \tag{7-31}$$

As the sequence of substitutions is continued, we have in the limit

$$x_n = (1 - e^{-T/\tau}) (Q_{n-1} + e^{-T/\tau} Q_{n-2} + e^{-2T/\tau} Q_{n-3} + \cdots) \tag{7-32}$$

If this series is truncated, we call it the finite memory form of the original recursion formula (infinite memory form) for x_n.

Now, to approximately compute the response of a continuous process to a random variable input, the mean squared value of the output is calculated as a function of the mean square value of the input in the following manner. Assume

$$x_n \cong \sum_i a_i Q_{n-i} \tag{7-33}$$

Then

$$x_n^2 \approx \sum_i \sum_j a_i a_j Q_{n-i} Q_{n-j} \tag{7-34}$$

$$\overline{x_n^2} = \sigma_x^2 \cong \sum_i a_i^2 \,\overline{Q_i^2} = (\Sigma a_i^2)\sigma_Q^2 \tag{7-35}$$

provided that $Q(t)$ is stationary and "almost white." In our example we can write

$$x_n = (1 - e^{-T/\tau})(Q_{n-1} + e^{-T/\tau}Q_{n-2} + \cdots) \tag{7-36}$$

$$x_n^2 = (1 - e^{-T/\tau})^2(Q_{n-1}^2 + e^{-2T/\tau}Q_{n-2}^2 + \cdots) \tag{7-37}$$

$$\overline{x_n^2} = (1 - e^{-T/\tau})^2\left(\overline{Q_{n-1}^2} + e^{-2T/\tau}\,\overline{Q_{n-2}^2} + \cdots\right) \tag{7-38}$$

$$\sigma_x^2 = (1 - e^{-T/\tau})^2(1 + e^{-2T/\tau} + e^{-4T/\tau} + \cdots)\sigma_Q^2 \tag{7-39}$$

$$\sigma_x^2 = \frac{(1 - e^{-T/\tau})^2}{(1 - e^{-2T/\tau})}\sigma_Q^2 \tag{7-40}$$

In general, then, the variance transfer function from input to output for almost-white and stationary noise inputs is given by

$$\frac{\sigma_x^2}{\sigma_Q^2} = \Sigma a_i^2 \tag{7-41}$$

It is clear from these equations that if the process being simulated is unstable, the propagation of roundoff and truncation error is also unstable. If the process being simulated is stable, however, the roundoff and truncation error will eventually reach the equilibrium condition set by the variance transfer function. All that remains is to compute the variance of the roundoff or truncation error, which can be done by other means covered extensively in others books on numerical analysis. Suffice it to say that if part of the task is to analyze the continuous system's response to noise, these variance propagation transfer functions can be used to approximately predict the continuous system's response to noise input.

7-5 REFERENCES

For this chapter refer to J. M. Smith, "Recent Developments in Numerical Integration," *ASME Journal of Dynamics, Measurement, and Control*, March 1974, pages 61–67, and M. E. Fowler, "A New Numerical Method for Simulation," *Simulation*, Vol. 4, May 1965, pages 324–330.

CHAPTER 8

CHEBYSHEV AND RATIONAL POLYNOMIAL APPROXIMATIONS FOR ANALYTIC SUBSTITUTION

8-1 INTRODUCTION

In this chapter we are concerned not so much with the numerical evaluation of functions or analyzing data as with deriving polynomials that can be used for analytic substitution. Existing handbooks often give the series expansion of many advanced functions that, though useful for analytical work, converge too slowly to serve in numerical analysis with the pocket calculator. These series can be modified to converge more quickly using Chebyshev polynomials or rational polynomial approximations. Polynomial approximations for truncated series expansions of functions for pocket calculator analysis have the advantage that they can be written in nested parenthetical form and efficiently evaluated on the pocket calculator with high precision. The objective here, then, is to improve the convergence of series approximations of given functions.

The Chebyshev polynomials can be used in a unique process. commonly called economization, to transform a truncated power series expansion of a function into a more quickly converging polynomial. They can, therefore, transform tables of infinite series of questionable value for numerical analysis into fast converging series (of which the error is well known) by the Chebyshev approximation theorem. Again, these polynomials can be written in nested parenthetical form for pocket calculator evaluation. In short, Chebyshev polynomials make tables of infinite series representation of advanced mathematical functions (of which there are a great many) eminently practical for pocket calculator evaluation. Because economization is so easy to perform and makes large tables of infinite series

198

immediately available, this chapter is dedicated to instructing the analyst on the marvelous properties of Chebyshev polynomials and their application to conditioning series for improved convergence.

Chebyshev polynomials have five important mathematical properties:

1. They are orthogonal polynomials, with a suitable weighting function, whether defined on a continuous interval or a discrete set of points.

2. They are equal-ripple functions; that is, they alternate between maxima and minima of the same size.

3. The zeros of successive Chebyshev polynomials interlace each other.

4. All Chebyshev polynomials satisfy a three-term recurrence relation.

5. They are easy to compute and to convert to and from a power series form.

These properties together generate an approximating polynomial which minimizes the maximum error in its application. This is quite different from, for example, least squares approximation where the sum of the squares of the errors is minimized. In least squares approximations the average square error is minimized; the maximum error itself can be quite large. In the Chebyshev approximation, the average error can be large but the maximum error is minimized. Chebyshev approximations of a function are sometimes said to be *mini-max* approximations of the function.

8-2 CHEBYSHEV POLYNOMIALS DEFINED

The Chebyshev polynomials are simply defined by the relations

$$T_0(x) = 1 \tag{8-1}$$

$$T_n(x) = \cos(n\theta) \tag{8-2}$$

$$\cos\theta = x \tag{8-3}$$

Equation 8-2 shows that the Chebyshev polynomials are orthogonal (with a suitable weighting factor), since cosine is an orthogonal function and cos $(n\theta)$ is a polynomial of degree n in $\cos\theta$.

Noting the trigonometric identity

$$\cos(n+1)\theta + \cos(n-1)\theta = 2\cos\theta\cos n\theta \tag{8-4}$$

we can write immediately that

$$T_{n+1} + T_{n-1} = 2x\, T_n \tag{8-5}$$

$$\therefore T_{n+1} = 2xT_n - T_{n-1} \tag{8-6}$$

Using this recurrence relation for the Chebyshev polynomials we can easily generate the successive polynomials as follows: Since

$$T_0 = 1$$

and

$$T_1 = x$$

in equation 8-6 we find

$$T_2 = 2xT_1 - T_0 = 2x^2 - 1$$

Then starting with

$$T_1 = x$$

and

$$T_2 = 2x^2 - 1$$

and again using the recurrence formula 8-6, we find

$$T_3 = 2xT_2 - T_1 = 4x^3 - 3x$$

Continuing in a similar manner we can form the table of Chebyshev polynomials:

$$T_0 = 1$$

$$T_1 = x$$

$$T_2 = 2x^2 - 1$$

$$T_3 = 4x^3 - 3x$$

$$T_4 = 8x^4 - 8x^2 + 1$$

$$T_5 = 16x^5 - 20x^3 + 5x$$

$$T_6 = 32x^6 - 48x^4 + 18x^2 - 1$$

$$T_7 = 64x^7 - 112x^5 + 56x^3 - 7x$$

$$T_8 = 128x^8 - 256x^6 + 160x^4 - 32x^2 + 1$$

Note that we can also form a table for powers of x in terms of

Chebyshev polynomials by simply solving for the powers of x from this table:

$$1 = T_0$$

$$x = T_1$$

$$x^2 = \frac{T_0 + T_2}{2}$$

$$x^3 = \frac{3T_1 + T_3}{4}$$

$$x^4 = \frac{3T_0 + 4T_2 + T_4}{8}$$

$$x^5 = \frac{10T_1 + 5T_3 + T_5}{16}$$

$$x^6 = \frac{10T_0 + 15T_2 + 6T_4 + T_6}{32}$$

$$x^7 = \frac{35T_1 + 21T_3 + 7T_5 + T_7}{64}$$

$$x^8 = \frac{35T_0 + 56T_2 + 28T_4 + 8T_6 + T_8}{128}$$

What is important about the Chebyshev polynomials is that Chebyshev proved that of all polynomials of degree n having a leading coefficient of 1, these polynomials (when divided by 2^{n-1}) have the least extreme value in the interval

$$-1 \leqslant x \leqslant +1$$

No other polynomials of degree n, whose leading coefficient is 1, have a smaller extreme value than

$$\max \left| \frac{T_n(x)}{2^{n-1}} \right| = \frac{1}{2^{n-1}}$$

in the interval $-1 \leqslant x \leqslant +1$ since $\max(|T_n(x)| = |\cos n\theta|) = 1$. This is an extremely important finding because it says that if we approximate a function in the interval $|x| \leqslant 1$ with Chebyshev polynomials that are truncated at n terms, the maximum error in the approximation is

$$\frac{1}{2^{n-1}}$$

The objective, then, is to find an expansion for function $f(x)$ in terms of Chebyshev polynomials:

$$f(x) = \sum_{n=0}^{n} a_n T_n(x) \tag{8-7}$$

The error properties associated with the Chebyshev polynomials are so significant that we take a few moments to show heuristically that there are no other polynomials with these properties.

First note that the leading coefficient in the Chebyshev polynomials generated with the recurrence formulas for an nth-order polynomial is 2^{n-1}. Thus

$$\frac{T_n(x)}{2^{n-1}}$$

is a polynomial with leading coefficient 1. Also, since $T_n(x)$ is a cosine function ($x = \cos\theta$), in the interval

$$0 \leqslant \theta \leqslant \pi$$

there are $n+1$ maxima alternating from $+1$ to -1. Clearly our normalized Chebyshev polynomial also has $n+1$ extreme values on the interval.

To prove that there is no other nth-order polynomial with leading coefficient 1 which has smaller extreme values in the interval, we assume that there is such a polynomial and prove that it must be a Chebyshev polynomial.

Assume that there is a polynomial $c(x)$ of degree n with leading coefficient 1 which has a smaller extreme value in the interval than the extremes of our normalized Chebyshev polynomial. Then

$$\frac{T_n(x)}{2^{n-1}} - c(x) = J(x)$$

is a polynomial that has $n+1$ maxima alternating in sign n times in the interval $|x| \leqslant 1$; thus $J(x)$ has n roots. But the polynomial formed by the difference between the normalized Chebyshev polynomial and the polynomial $c(x)$ is of degree $(n-1)$. Thus a polynomial of degree $(n-1)$ can have n zeros only if the polynomial is zero. If

$$J(x) = 0 = \frac{T_n(x)}{2^{n-1}} - c(x)$$

then

$$c(x) = \frac{T_n(x)}{2^{n-1}}$$

Because of the power of the Chebyshev polynomial approximation, its orthogonality properties are worth examining to make sure that the Chebyshev polynomials are orthogonal and that a formula can be developed for deriving the coefficients in the series expansion of equation 8-7. We can determine the orthogonality properties or characteristics for the Chebyshev polynomials from what we know of the orthogonality of the cosine functions: that is,

$$\int_0^\pi \cos(m\theta)\cos(n\theta)\,d\theta \begin{cases} 0 & (m \neq n) \\ \pi/2 & (m = n \neq 0) \\ \pi & (m = n = 0) \end{cases}$$

Then, substituting

$$T_n(x) = \cos(n\theta)$$

$$\cos\theta = x$$

to obtain the orthogonality properties of the Chebyshev polynomials, we find

$$\int_{-1}^{1} \frac{T_m(x)T_n(x)}{(1-x^2)^{1/2}}\,dx = \begin{cases} 0 & (m \neq n) \\ \pi/2 & (m = n \neq 0) \\ \pi & (m = n = 0) \end{cases} \tag{8-8}$$

We see, then, that the Chebyshev polynomials form an orthogonal set on the interval

$$-1 \leqslant x \leqslant +1$$

with the weighting function

$$w(x) = \frac{1}{(1-x^2)^{1/2}}$$

The Chebyshev polynomials can also be shown to be orthogonal over a *discrete* set of x_n. These orthogonality conditions can then be used to evaluate the coefficients a_n in equation 8-7:

$$a_n = \frac{2}{\pi} \int_{-1}^{+1} \frac{f(x)T_n(x)}{(1-x^2)^{1/2}} dx, \quad (n \geqslant 1) \tag{8-9}$$

$$a_0 = \frac{1}{\pi} \int_{-1}^{+1} \frac{f(x)}{(1-x^2)^{1/2}} dx, \quad (n=0) \tag{8-10}$$

Similarly, if the function $f(x)$ is only defined on

$$x_p = \cos \frac{\pi}{N} (p + \tfrac{1}{2})$$

then the coefficients in the expansion

$$f(x) = \sum_{n=0}^{N-1} a_n T_n(x)$$

are given by

$$a_n = \frac{2}{N} \sum_{p=0}^{N-1} f(x_p) T_n(x_p)$$

$$a_0 = \frac{1}{N} \sum_{p=0}^{N-1} f(x_p)$$

for the discrete (which is easier to use) or the continuous method for preparing a Chebyshev approximation of a function $f(x)$.

There is another approach to approximating $f(x)$ with Chebyshev polynomials that is due primarily to Lanczos. It is powerful and simple to use, having the nice property that it will usually improve the convergence of any truncated series expansion of a function $f(x)$. We do the following:

1. Write a truncated series or polynomial approximation of $f(x)$ in nested form.
2. Rewrite the polynomial in terms of Chebyshev polynomials.
3. Truncate the Chebyshev approximation by an additional one or two terms.
4. Rewrite the Chebyshev polynomials in terms of polynomials in x.
5. Rewrite this polynomial in nested parenthetical form for convenient numerical evaluation on the pocket calculator.

For example, if we write a truncated power series representation of a function in the form

$$f(x) = \sum_{n=0}^{m} a_n x^n \tag{8-11}$$

by rewriting equation 8-11 in the nested parenthetical form

$$f(x) = a_0 + x(a_1 + x(a_2 + \cdots + x(a_{m-1} + a_m x) \cdots))$$

we can convert this to a series of Chebyshev polynomials by starting at the inner parentheses and rewriting it in the form

$$a_{n-1} + a_m x = a_{m-1} T_0 + a_m T_1 \tag{8-12}$$

Assuming that at the nth parenthetical nest

$$\alpha_0 T_0 + \alpha_1 T_1 + \cdots + \alpha_n T_n \tag{8-13}$$

we can multiply this by x and add to it the next coefficient in the power series a_{M-n-1} to get the $(n+1)$st nest. Then, by using the relationships

$$xT_0 = T_1$$

$$xT_n = \frac{T_{n+1} + T_{n-1}}{2} \tag{8-14}$$

the power series in the nth parentheses is transformed in the $(n+1)$st power series in Chebyshev polynomials as

$$\frac{x_n T_n}{2} + \frac{\alpha_{n-1} T_{n-1}}{2} + \cdots + \left(\frac{\alpha_1 + \alpha_3}{2}\right) T_2 + \left(\alpha_0 + \frac{\alpha_2}{2}\right) T_1 + \left(\alpha_{M-n-1} + \frac{a_1}{2}\right) T_0$$

$$\tag{8-15}$$

The process for generating the coefficients at any given stage in the development of the Chebyshev polynomial expansion of $F(x)$ can be visualized as shown in Figure 8-11. Here the coefficients at a given stage are used to generate the coefficients in the next stage according to the diagram.

In this approach, the coefficient associated with the Mth term of the original series

$$a_m x^m$$

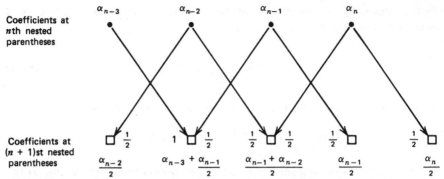

Figure 8-1   Process for generating coefficients for the Chebyshev expansion of $f(x)$.

becomes, in the Chebyshev polynomial expansion,

$$\frac{a_m}{2^{m-1}}T_m(x)$$

Thus if we truncated the Chebyshev polynomial expansion of $f(x)$ beginning with the mth term, the error would be on the order of $[a_m/(2^{m-1})]$ instead of a_m as in the original polynomial approximation (see Table 8-1).

Table 8-1   Expansions of $f_N(x) = \sum\limits_{n=0}^{N} a_n x^n$ in Chebyshev Polynomials

Expansion in powers of x	Chebyshev expansion
$f_0 = a_0$	$a_0 T_0$
$f_1 = a_0 + a_1 x$	$a_0 T_0 + a_1 T_1$
$f_2 = a_0 + a_1 x + a_2 x^2$	$\left(a_0 + \dfrac{a_2}{2}\right)T_0 + a_1 T_1 + \left(\dfrac{a_2}{2}\right)T_2$
$f_3 = a_0 + a_1 x + a_2 x^2 + a_3 x^3$	$\left(a_0 + \dfrac{a_2}{2}\right)T_0 + \left(a_1 + \dfrac{3a_3}{4}\right)T_1 + \left(\dfrac{a_2}{2}\right)T_2 + \left(\dfrac{a_3}{4}\right)T_3$
$f_4 = a_0 + a_1 x + a_2 x^2 + a_3 x^3 + a_4 x^4$	$\left(a_0 + \dfrac{a_2}{2} - \dfrac{a_4}{4}\right)T_0 + \left(a_1 + \dfrac{3a_3}{4}\right)T_1 + \tfrac{1}{2}(a_2 + a_4)T_2 + \left(\dfrac{a_3}{4}\right)T_3 + \left(\dfrac{a_4}{8}\right)T_4$
$f_5 = a_0 + a_1 x + a_2 x^2 + a_3 x^3 + a_4 x^4$ $\quad + a_5 x^5$	$\left(a_0 + \dfrac{a_2}{2} - \dfrac{a_4}{2}\right)T_0 + \left(a_1 + \dfrac{3a_3}{4} + \dfrac{10a_5}{16}\right)T_1 + \tfrac{1}{2}\left(\dfrac{a_2 + a_4}{2}\right)T_2 + \left(\dfrac{a_3}{4} + \dfrac{5a_5}{16}\right)T_3 + \left(\dfrac{a_4}{8}\right)T_4 + \left(\dfrac{a_5}{16}\right)T_5$

In this sense, then, we say that the Chebyshev expansion converges more quickly than does the original expansion.

Another way of looking at it is as follows. If the original polynomial approximation of the function was accurate to some error, the Chebyshev polynomial approximation will usually be almost as accurate to the same error with fewer terms. Additional terms can therefore be dropped from the reexpressed polynomial. This is the process called economization.

The procedure followed in equations 8-12 through 8-15 shows the approach to generating Chebyshev polynomial approximations of $f(x)$ in general for n terms. However, polynomials of a degree higher than 5 are generally cumbersome for pocket calculator analysis. The table of powers of x in terms of the Chebyshev polynomials, presented earlier in the chapter and partly repeated here for convenience, is useful for direct substitution of powers of x for its Chebyshev polynomial equivalent.

$$1 = T_0$$

$$x = T_1$$

$$x^2 = \frac{T_0 + T_2}{2}$$

$$x^3 = \frac{3T_1 + T_3}{4}$$

$$x^4 = \frac{3T_0 + 4T_2 + T_4}{8}$$

$$x^5 = \frac{10T_1 + 5T_3 + T_5}{16}$$

In this way a power series such as

$$f(x) = a_0 + a_1 x + a_2 x^2 + \cdots + a_M x^M$$

can be converted into an expansion in Chebyshev polynomials:

$$f(x) = b_0 + b_1 T_1 + b_2 T_2 + \cdots + b_M T_M$$

For this process to be workable the series must be written in a form where the evaluation of $f(x)$ takes place for x on the interval $(-1 < x < +1)$. Once the expansion is written in terms of Chebyshev polynomials, they can be replaced by polynomials in x from our earlier table, repeated here for convenience:

$$T_0 = 1$$

$$T_1 = x$$

$$T_2 = 2x^2 - 1$$

$$T_3 = 4x^3 - 3x$$

$$T_4 = 8x^4 - 8x^2 + 1$$

$$T_5 = 16x^5 - 20x^3 + 5x$$

They can then be algebraically simplified and written in nested parenthetical form for quick evaluation on the pocket calculator.

Hamming works the easy-to-follow example

$$y = \ln(1 + x) \cong x - \frac{x^2}{2} + \frac{x^3}{3} * \tag{8-16}$$

to illustrate the method of economization. By direct substitution for powers of x we can rewrite this power series expansion in terms of the Chebyshev polynomials as

$$y \cong T_1 - \left(\frac{T_0 + T_2}{4} \right) + \left(\frac{3T_1 + T_3}{12} \right)$$

$$y \cong -\frac{T_0}{4} + \left(1 + \frac{3}{12} \right) T_1 - \frac{T_2}{4} + \frac{T_3}{12} \tag{8-17}$$

$$y \cong -\frac{T_0}{4} + \frac{15 T_1}{12} - \frac{T_2}{4} + \frac{T_3}{12}$$

Dropping the last term in the power series (equation 8-16) results in dropping 0.25 from the numerical evaluation of y (when $x = 1$); a roughly equivalent error is produced in the power series (equation 8-17) when dropping the last two terms. This can be seen by noting that at most the error will be

$$\epsilon = \tfrac{1}{12} - \tfrac{1}{4} = -0.1766 \cdots$$

---

*Hamming carries five terms—more than we need to illustrate the principle for pocket calculator polynomials of convenient size.

We know this because the magnitude of the value of the Chebyshev polynomials is less than or equal to 1 for all x on the interval. Thus we can write

$$y = \ln(1+x) \cong -\frac{T_0}{4} + \frac{15 T_1}{12} - \frac{T_2}{4} \tag{8-18}$$

with somewhat better accuracy than is given by equation 8-16. This, then, is the process of economization. Using the definitions of the Chebyshev polynomials we can rewrite equation 8-18 in the form

$$y = \ln(1+x) \cong -\frac{1}{4} + \frac{15x}{12} - \frac{2x^2}{4} + \frac{1}{4} = x(1.25 - 0.5x)$$

The numerical comparison of the economized second-order and the non-economized third-order polynomials is given below. The economized quadratic equation has smaller average error (-0.0676) than the non-economized cubic equation (0.1993); it also has the smallest maximum error $(0.1$ at $x = 0.5)$ on the interval

$$0 \leqslant x \leqslant 1$$

x	Exact $y = \ln(1+x)$	Economized $y \cong x(1.25 - 0.5x)$		Noneconomized $y \cong x - 0.5x^2 + 0.333x^3$	
		y	Error	y	Error
0.1	0.09531018	0.12000000	-0.0246	0.0948333	0.0052
0.3	0.26236426	0.33000000	-0.0676	0.25050000	0.0495
0.6	0.47000363	0.57000000	-0.0999	0.38400000	0.2160
0.9	0.64185389	0.72000000	-0.0781	0.37350000	0.5262

Let us recap Chebyshev polynomials in the context of pocket calculator analysis. We saw in Chapter 1 that the evaluation of polynomials greater than third or fourth-order involved a sizable number of key strokes on the pocket calculator. We found that by writing these polynomials in nested parenthetical form we could go to fifth and sixth-order polynomials with the same number of key strokes as required for the third-order polynomial evaluation. This gave us additional accuracy with the same number of key strokes. Now we have found that the economization process due to Chebyshev polynomials can occasionally provide high-order polynomial accuracy with low-order polynomials, even further reducing the number of key strokes for, say, equivalent fifth-order polynomial evaluation accuracy. That is, retaining fifth-order polynomial accuracy with Chebyshev polynomials can provide equivalent accuracy of seventh- or eighth-order noneconomized polynomial expansions. Then when it is written in nested parenthetical form, the Chebyshev polynomial provides this accuracy with many fewer key strokes than would normally be required for up to eighth- or ninth-order polynomial expansions. Thus the nested parenthetical evaluation of Chebyshev polynomial approximations reduces the *workload* on the pocket calculator from that associated with a ninth-order polynomial approximation of $f(x)$ to that of a second- or third-order polynomial approximation of $f(x)$. This results in an order-of-magnitude reduction in key strokes.

In general, the approach to evaluating advanced mathematical functions is (1) to write the function in a truncated polynomial form, (2) rewrite that expression so that the interval on which $f(x)$ is to be evaluated is between $-1 \leqslant x \leqslant +1$, (3) economize the series using Chebyshev polynomials, (4) rewrite the Chebyshev approximation to the function in nested parenthetical form, and (5) use it for numerical evaluation on the pocket calculator.

Numerical Evaluation of Chebyshev Polynomials

It is useful to know that the recurrence formula (recopied here for convenience) for the Chebyshev polynomial can be used to numerically evaluate Chebyshev polynomials:

$$T_n(x) = 2xT_{n-1}(x) - T_{n-2}(x)$$

The starting values for the recurrence formula can be computed with:

$$T_0 = 1$$

$$T_1 = x$$

Thus the Chebyshev polynomial expansion of a function, once written, need not necessarily be given in powers of x but can be numerically evaluated directly. For example, the equation

$$y = \ln(1+x) \cong -\frac{T_0}{4} + \frac{15}{12}T_1 - \frac{T_2}{4} + \frac{T_3}{12} \tag{8-17}$$

is the Chebyshev approximation to $\ln(1+x)$, which was developed earlier. Using the recursion formula and the fact that $T_0 = 1$ and $T_1 = x$, we can now numerically evaluate equation 8-17 by first evaluating the numerical values for the five Chebyshev polynomials; for example, when $x = 0.3$,

$$T_0 = 1$$

$$T_1 = 0.3$$

$$T_2 = (2)(0.3)T_1 - T_0 = 2 \times 0.3 \times 0.3 - 1 = -0.82$$

$$T_3 = (2)(0.3)T_2 - T_1 = -2 \times 0.3 \times 0.82 - 0.3 = -0.792$$

These can be substituted in the power series expansion to numerically evaluate the series:

$$y = \ln(1.3) \cong -\frac{1}{4} + \frac{15}{12} \times 0.3 - \frac{1}{4}(-0.82) + \frac{1}{12}(-0.792) = 0.2640$$

$$y = \ln(1.3) \equiv 0.26236426$$

This procedure allows convenient numerical evaluation of high-order Chebyshev polynomials (e.g., 20). Although writing the nested parenthetical form of the Chebyshev polynomial expansion of a function is possible, it is cumbersome and can trip-up the user if he forgets which parentheses he is at in the numerical evaluation process. A better alternative is to compute first the numerical values of the Chebyshev polynomials and then substitute them into the polynomial expansion equation, since this does not directly involve the evaluation of high-order polynomials.

8-3 APPROXIMATION FOR ANALYTIC SUBSTITUTION WITH RATIONAL POLYNOMIALS

We have seen so far that the expansion of a function in terms of Chebyshev polynomials can be used to series expand a function that minimizes the maximum error in the approximation on the interval $-1 \leqslant x \leqslant +1$. As noted in Chapter 3, the economized Chebyshev approximations were not extensively employed in precision evaluation of functions. The reason is that they are not necessarily the best approximation for pocket calculators—from the standpoint of the time required to evaluate the function and the storage needed to store the coefficients. A better approach is to use the ratio of two polynomials as a means for approximating functions. Again, this is so because nested polynomials are conveniently numerically evaluated on any digital computer including the pocket calculator.

Consider the case where we wish to represent a function as the quotient of two polynomials:

$$f(x) \cong \frac{a_0 + a_1 x + a_2 x^2 + \cdots + a_n x^n}{1 + b_1 x + b_2 x^2 + \cdots + b_m x^m} = R_N(x), \qquad (N = n + m)$$

The rational polynomial approximations used here are those whose numerator is equal to or greater by 1 than the degree of the denominator. The constant term in the denominator can be taken to be 1 without loss of generality. Since our concern here is the interval $-1 \leqslant x \leqslant +1$, we can use the Maclaurin series expansion for $f(x)$ as a means of determining the coefficients in the rational polynomial approximation. The number of terms that we would use in the Maclaurin series is equal to the sum of the order in the numerator and denominator, because this is the number of coefficients that must be determined. If we write

$$f(x) = (f_0 + c_1 x + c_2 x^2 + \cdots + c_n s^n)$$

for the Maclaurin series expansion of $f(x)$, the difference between the Maclaurin series and the rational polynomial approximation can be formed as follows:

$$f(x) - R_N(x) = (c_0 + c_1 x + c_2 x^2 + \cdots + c_N x^N) - \frac{a_0 + a_1 x + \cdots + a_n x^n}{1 + b_1 x + \cdots + b_m x^m}$$

$$f(x) - R_N(x) = \frac{(c_0 + c_1 x + c_2 x^2 + \cdots + c_N x^N)(1 + b_1 x + \cdots + b_m x^m)}{1 + b_1 x + \cdots + b_m x^m}$$

$$\tag{8-19}$$

Now, if

$$f(x) = R_N(x) \text{ at } x = 0$$

then

$$c_0 - a_0 = 0$$

Similarly, for the first N derivatives of $f(x)$ and $R_N(x)$ to be equal at $x = 0$ the coefficients of the powers of x in the numerator must all be zero. This gives the system of equations shown below:

$$b_1 c_0 + c_1 - a_1 = 0$$

$$b_2 c_0 + b_1 c_1 + c_2 - a_2 = 0$$

$$b_3 c_0 + b_2 c_1 + b_1 c_2 + c_3 - a_3 = 0$$

$$\vdots$$

$$b_m c_{n-m} + b_{m-1} c_{n-m+1} + \cdots + c_n - a_n = 0$$

$$b_m c_{n-m+1} + b_{n-1} c_{n-m+2} + \cdots + c_{n+1} = 0$$

$$b_m c_{n-m+2} + b_{m-1} c_{n-m+3} + \cdots + c_{n+2} = 0$$

$$\vdots$$

$$b_m c_{n-m} + b_{m-1} c_{n-m+1} + \cdots + c_n = 0$$

When combined with equation 8-19, this can be solved for all coefficients of the rational polynomial.

The process just described is that of forming the Padé approximation. We illustrate Padé approximations with a simple example.

Example. Consider approximating $\sin(x)$ with a rational polynomial where $N = 9$. We use a fifth-degree polynomial in the numerator. The Maclaurin series expansion through x^9 for arctan (x) is

$$\arctan(x) \cong x - \frac{x^3}{3} + \frac{x^5}{5} - \frac{x^7}{7} + \frac{x^9}{9} = M_9(x)$$

Following the procedure outlined above, we form the equation

$$f(x) - R_9(x) = \frac{(1 - \frac{1}{3}x^3 + \frac{1}{5}x^5 - \frac{1}{7}x^7 + \frac{1}{9}x^9)(1 + b_1 x + \cdots + b_4 x^4)}{1 + b_1 x + b_2 x^2 + \cdots + b_4 x^4} \\ {- (a_0 + \cdots + a_5 x^5)}$$

from which we can evaluate the coefficients for the rational polynomial:

$$a_0 = 0 \qquad\qquad b_1 = \tfrac{5}{3} b_3$$

$$a_1 = 1 \qquad\qquad b_2 = \frac{5}{7} + \frac{5}{3} b_4$$

$$a_2 = b_1 \qquad\qquad b_3 = \tfrac{5}{7} b_1$$

$$a_3 = -\tfrac{1}{3} + b_2 \qquad b_4 = -\tfrac{5}{9} + \tfrac{5}{7} b_2$$

$$a_4 = -\tfrac{1}{3} b_1 + b_3$$

$$a_5 = \tfrac{1}{5} - \tfrac{1}{3} b_2 + b_4$$

Solving for the a's and b's, we find

$$a_0 = 0$$

$$a_1 = 1$$

$$a_2 = 0$$

$$a_3 = \tfrac{7}{9}$$

$$a_4 = 0$$

$$a_5 = \tfrac{64}{945}$$

$$b_1 = 0$$

$$b_2 = \tfrac{10}{9}$$

$$b_3 = 0$$

$$b_4 = \tfrac{5}{21}$$

Thus the rational polynomial for approximating arctan (x) is given by

$$\arctan(x) \cong \frac{x + \tfrac{7}{9}x^3 + \tfrac{64}{945}x^5}{1 + \tfrac{10}{9}x^2 + \tfrac{5}{21}x^4} = \frac{x\left(1 + \tfrac{7}{9}x^2\left(1 + \tfrac{576}{6615}x^2\right)\right)}{1 + \tfrac{10}{9}x^2\left(1 + \tfrac{45}{210}x^2\right)}$$

The table below compares the rational polynomial approximation (Padé approximation) and the Maclaurin series expansion for arctan (x).

x	Arctan (x)	$R_9(x)$	Error	$M_9(x)$	Error
0.2	0.19740	0.19740	0.00000	0.19740	0.00000
0.6	0.54042	0.54042	0.00000	0.54067	−0.00025
1.0	0.78540	0.78558	−0.00018	0.83492	−0.04952

Errors in Padé approximations can be estimated by computing the next nonzero term in the numerator of the rational polynomial. The procedure, though somewhat tedious, gives an error formula in terms of the next term with a nonzero coefficient. Furthermore, the error formula can be evaluated over the interval in which the function is being evaluated. The alternative is to work out a simple error curve as shown in our examples here; this can be conveniently done on the pocket calculator.

The rational polynomial approximation is significantly more accurate than the Maclaurin approximation. When $x = 1$, the Maclaurin series ceases to converge while the Padé approximation is still fairly accurate.

In the pocket calculator evaluation of the rational polynomial approximation we write the numerator and denominator in nested parenthetical form, numerically evaluate each separate polynomial, and then perform the division numerically.

Similarly we could have started with the Chebyshev series for arctan (x) on the interval $+1$ to -1, formed a rational polynomial in Chebyshev polynomials, and then proceeded to evaluate the coefficients in the approximation. The procedure differs from that above in that the product of the two Chebyshev polynomials results in squares and products of the polynomials themselves. In turns out, however, that the product of two Chebyshev polynomials can be given by

$$T_n(x)T_m(x) = \tfrac{1}{2}(T_{n+m}(x) + T_{n-m}(x))$$

here we have again a useful form for evaluating the coefficients in the numerator of the rational Chebyshev polynomial approximation formula. Once the approximation in Chebyshev polynomials is determined, it can be reduced to approximations in the independent variable on the interval $+1$ to -1.

As a simple example, consider the Chebyshev polynomial expansion of e^x given by (see Example 8-2 on page 220)

$$e^x \cong 1.2661 T_0 + 1.1303 T_1 + 0.2715 T_2 + 0.0444 T_3 = f(x)$$

Using the principles of rational polynomial approximation, we find the approximation R_3 in the form

$$R_3 = \frac{a_0 + a_1 T_1 + a_2 T_2}{1 + b_1 T_1}$$

Next we form the function $f(x) - R_3$. Setting the powers of x in the numerator equal to zero, we then find

$$f - R = \frac{(1.2661 + 1.1303 T_1 + 0.0444 T_3)(1 + b_1 T) - (a_0 + a_1 T_1 + a_2 T_2)}{1 + b_1 T_1}$$

of which the numerator becomes

$$1.2661 + 1.1306\,T_1 + 0.2715\,T_2 + 0.0444\,T_3 + 1.2661 b_1 T_1 + 1.1303 b_1 T_1^2$$

$$+ 0.2715 b_1 T_1 T_2 + 0.0444 b_1 T_1 T_3 - a_0 - a_1 T_1 - a_2 T_2$$

Remembering that

$$T_n(x) T_m(x) = \tfrac{1}{2}[T_{n+m}(x) + T_{n-m}(x)]$$

we can write (equating $f(x)$ to R_3 and equating their first three derivatives, respectively):

$$a_0 = 1.2661 + \frac{1.1303}{2} b_1$$

$$a_1 = 1.303 + \left(\frac{0.2715}{2} + 1.2661 \right) b_1$$

$$a_2 = 0.2715 + \left(\frac{1.1303}{2} + \frac{0.0444}{2} \right) b_1$$

$$0 = 0.0444 + \frac{0.2715}{2} b_1$$

Thus

$$b_1 = -0.3266$$

$$a_0 = 1.0815$$

$$a_1 = 0.6724$$

$$a_2 = 0.07966$$

with the result that

$$e^x \cong \frac{1.0815 + 0.6724\,T_1 + 0.07966\,T_2}{1 - 0.3266\,T_1}$$

Then substituting

$$T_1 = x$$

$$T_2 = 2x^2 - 1$$

we find

$$e^x = \frac{1.0018 + 0.6724 x + 0.1593 x^2}{1 - 0.3266 x}$$

A comparison of the two types of approximations is shown in Figure 8-2. It is apparent that the approximation found by

$$e^x = \left(\frac{R_3 + C_3}{2}\right) + \left(\frac{\epsilon_R + \epsilon_C}{2}\right)$$

(where C_3 is the Chebyshev approximation of $f(x)$ developed in example 8-2 on page 220) has the error curve shown as a dashed line in the figure and is somewhat more accurate than either approximation alone.

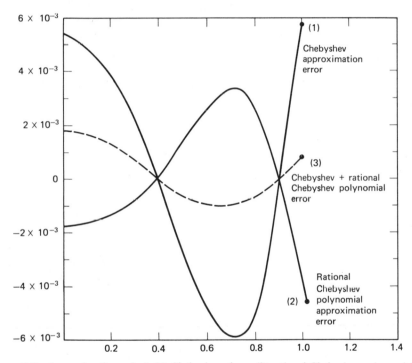

Figure 8-2   Approximations of e^x. (1) Chebyshev alone, (2) rational Chebyshev polynomials, and (3) a combination of both.

Example 8-1.   Prepare the rational polynomial approximation $\sin(x) = R_5(x)$ in the neighborhood of $x = 0$. Then

$$R_5(x) = \frac{a_0 + a_1 x + a_2 x^2 + a_3 x^3}{1 + b_1 x + b_2 x^2}$$

and

$$\sin(x)x - \frac{x^3}{6} + \frac{x^5}{120} \cong c_0 + c_1 x + c_2 x^2 + c_3 x^3 + c_4 x^4 + c_5 x^5$$

From R_5 and $\sin(x)$ we see that

$$
\begin{array}{lll}
c_0 = 0 & a_0 = a_0 & b_0 = 1 \\
c_1 = 1 & a_1 = a_1 & b_1 = b_1 \\
c_2 = 0 & a_2 = a_2 & b_2 = b_2 \\
c_3 = -\frac{1}{6} & a_3 = a_3 & b_3 = 0 \\
c_4 = 0 & a_4 = 0 & b_4 = 0 \\
c_5 = \frac{1}{120} & a_5 = 0 & b_5 = 0
\end{array}
$$

We know (see text) that the six equations for determining the six coefficients a_0, a_1, a_2, a_3, b_1, and b_2 are

$$c_0 - a_0 = 0$$

$$c_0 b_1 + c_1 - a = 0$$

$$c_0 b_2 + c_1 b_1 + c_2 - a_2 = 0$$

$$c_0 b_3 + c_1 b_2 + c_2 b_1 + c_3 - a_3 = 0$$

$$c_0 b_4 + c_1 b_3 + c_2 b_2 + c_3 b_1 + c_4 = 0$$

$$c_0 b_5 + c_1 b_4 + c_2 b_3 + c_3 b_2 + c_4 b_1 + c_5 = 0$$

On substituting directly from the preceding table of a's, b's, and c's, we find

$$a_0 = 0$$

$$a_1 = 1$$

$$a_2 = b_1$$

$$a_3 = -\tfrac{1}{6} + b_2$$

$$0 = -\frac{b_1}{6} \therefore b_1 = 0$$

$$0 = \tfrac{1}{120} - \frac{b_2}{6} \therefore b_2 = \tfrac{1}{20}$$

Substituting for b_2 in a_3 and b_1 in a_2, we find

$$a_0 = 0$$

$$a_1 = 1$$

$$a_2 = 0$$

$$a_3 = \left(\tfrac{1}{20} - \tfrac{1}{6}\right) = \left(\tfrac{6}{120} \quad \tfrac{20}{120}\right) = -\tfrac{14}{120}$$

$$b_1 = 0$$

$$b_2 = \tfrac{1}{20}$$

Thus

$$\sin(x) \cong \frac{x - \tfrac{14}{120}x^3}{1 + \tfrac{1}{20}x^2}$$

We can tabulate values of this approximation and compare them with the Maclaurin series approximation of $\sin(x)$ and the actual value of $\sin(x)$. This is done in Table 8-2.

Table 8-2

x	$\sin(x)$	R_5	Error	M_5	Error
0.25	0.24740396	0.24740395	1×10^{-8}	0.24740397	-1×10^{-8}
0.50	0.47942554	0.47942387	1.67×10^{-6}	0.47942708	-15.4×10^{-6}
0.75	0.68163876	0.68161094	27.82×10^{-6}	0.68166504	-26.28×10^{-6}
1.00	0.84147098	0.84126984	201.14×10^{-6}	0.84166667	-195.68×10^{-6}
1.25	0.74898462	0.94806763	916.99×10^{-6}	0.94991048	-925.86×10^{-6}
1.50	0.99749499	0.99438202	3112.96×10^{-6}	1.00078125	-3286.26×10^{-6}
$\pi/2$	1.00000000	0.99577290	4227.10×10^{-6}	1.00452486	-4524.86×10^{-6}
2.00	0.90929743	0.88888888	2.040854×10^{-2}	0.93333333	-2.403591×10^{-2}

In the interval between 0 and 1 the Maclaurin expansion is more accurate (as we might expect), while the rational polynomial tends to be somewhat more accurate outside the $(0, 1)$ interval.

This example makes the important point that rational polynomial expansions, though involving polynomials of lower powers than do the polynomial from which they were derived, are not necessarily equally accurate over all intervals. The procedure of generating rational polynomial approximations of functions is, however, also illustrated in the example.

Example 8-2.   Economize the Maclaurin series expansion of e^x:

$$e^x = 1 + x + \frac{x^2}{2} + \frac{x^3}{6} + \frac{x^4}{24} + \frac{x^5}{120} + \frac{x^6}{720} + \cdots$$

Since

$$1 = T_0$$

$$x = T_1$$

$$x^2 = \tfrac{1}{2}(T_0 + T_2)$$

$$x^3 = \tfrac{1}{4}(3T_1 + T_3)$$

$$x^4 = \tfrac{1}{8}(3T_0 + 4T_2 + T_4)$$

$$x^5 = \tfrac{1}{16}(10T_1 + 5T_3 + T_5)$$

$$x^6 = \tfrac{1}{32}(10T_0 + 15T_2 + 6T_4 + T_6)$$

we can rewrite e^x as

$$e^x = T_0 + T_1 + \tfrac{1}{4}(T_0 + T_2) + \tfrac{1}{24}(3T_1 + T_3) + \tfrac{1}{192}(3T_0 + 4T_2 + T_4)$$

$$+ \tfrac{1}{1920}(10T_1 + 5T_3 + \cdots) + \tfrac{1}{23040}(10T_0 + 15T_2 + \cdots) + \cdots$$

$$e^x = 1.2661 T_0 + 1.1303 T_1 + 0.2715 T_2 + 0.0444 T_3 + \cdots$$

$$e^x = 1.2661 + 1.1303x + 0.2715(2x^2 - 1) + 0.0444(4x^3 - 3x) + \cdots$$

$$e^x \cong 0.9946 + 0.9971x + 0.5430x^2 + 0.1776x^3 + \cdots$$

Note that the terms involving T_0, T_1, T_2 and T_3 were carried from substitutions of polynomials for up to six terms in the Maclaurin series. Thus we also continue the effect of the sixth term on the first and succeeding terms—the effect that makes the economization work.

Comparison of the Chebyshev and Maclaurin approximations is shown in Table 8-3.

Table 8-3

x	e^x	Maclaurin	Chebyshev	Error M	Error C
1.0	2.7183	2.6667	2.7123	0.0516	0.0060
0.8	2.2255	2.2053	2.2307	0.0202	−0.0052
0.6	1.8221	1.8160	1.8267	0.0061	−0.0057
0.4	1.4918	1.4907	1.4917	0.0011	0.0001
0.2	1.2214	1.2213	1.2172	0.0001	0.0042
0.0	1.0000	1.0000	0.9946	0.0000	0.0054

Notice that the Chebyshev error is a maximum at $x=0$ and the Maclaurin error is a minimum. This is because of the osculating nature of Maclaurin approximation at the origin as compared with the mini-max nature of the Chebyshev approximation on the interval $(0, 1)$. This is illustrated in Figure 8-3.

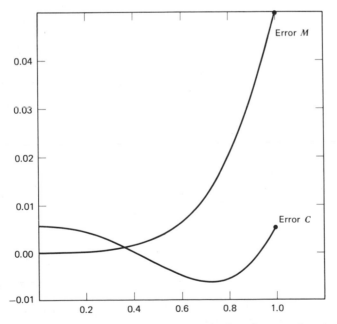

Figure 8-3 Error in Chebyshev economization of a Maclaurin series expansion of e^x.

Example 8-3.   Approximate $\sin(x)$ with Chebyshev polynomials by the method of economization.

In this simple example, we use the Maclaurin approximation of $\sin(x)$, again because it is an approximation centered on the interval $-1 \leqslant x \leqslant +1$. We see that

$$\sin(x) \cong x - \frac{x^3}{6} + \frac{x^5}{120}$$

Then

$$\sin x \cong T_1 - \tfrac{1}{24}(3T_1 + T_3) + \tfrac{1}{1920}(10T_1 + 5T_3 + T_5)$$

$$\sin x \cong \tfrac{169}{192}T_1 - \tfrac{5}{128}T_3 + \tfrac{1}{1920}T_5$$

The higher powers of x from the Maclaurin series would make further contributions to the T_1, T_3, and T_5 coefficients. The contributions are small, however, especially for the early T_x terms. The x^5 term, in particu-

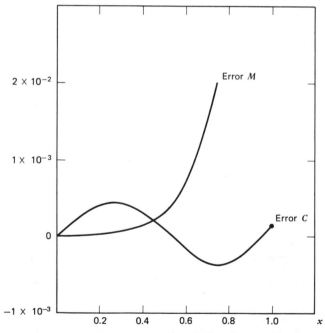

Figure 8-4 A comparison between the error that results from a Maclaurin approximation and an economized Chebyshev approximation of $\sin(x)$.

lar, changes the T_1 coefficient by less than 1% and the x^7 term alters it by less than 0.01%.

Economizing the Chebyshev approximation (dropping the T_5 term), we find

$$\sin(x) \cong \tfrac{169}{192} T_1 - \tfrac{5}{128} T_3$$

On substituting

$$T_1 = x$$

$$T_3 = 4x^3 - 3x$$

we find

$$\sin(x) \cong 0.9974x - 0.1562x^3$$

$$\sin(x) \cong x(0.9974 - 0.1562x^2)$$

The errors for the Maclaurin approximation of $\sin(x)$ and the economized Chebyshev approximation are compared in Figure 8-4.

8-4 REFERENCES

For this chapter consult Richard Hamming's *Numerical Methods for Scientists and Engineers* (McGraw-Hill, 1973), Chapters 28, 29, and 30. For further reading on some of the examples given in this chapter read Curtis F. Gerald's *Applied Numerical Analysis* (Addison-Wesley, 1970). The author has used Gerald's textbook in courses on numerical analysis and has found it to be detailed and easy to read.

CHAPTER 9

DETERMINING THE ROOTS
OF A FUNCTION

9-1 INTRODUCTION

In this chapter we examine in more detail the problem of finding the roots of a function, which we encountered briefly in Chapter 3. First we discuss the real zeros of continuous functions and then touch on the problem of complex zeros. The need to find roots of a continuous function arises frequently in engineering, usually when solving implicit equations, determining maxima and minima, or finding solutions of simultaneous equations. The methods particularly suited to pocket calculator analysis are considered here. They differ from the standard methods for evaluating roots on large-scale digital computers in that the analyst must understand the function whose roots he is trying to find so as to select the proper approach to the problem.

Three methods for finding the real roots of a function are discussed. Perhaps the most straightforward approach is that of bisecting the interval over which the root is expected to be identified. The bisection method is slowly converging, but is virtually foolproof in its application (i.e., it is almost impossible to misuse the method).

Another approach is a modified form of the "false-position" method developed by Hamming. It is a fast-converging method, but involves slightly more functional evaluations than does the bisection method. Furthermore, for certain functions the false-position method can converge more slowly than does the bisection method, but these functions are not frequently encountered in practical engineering analysis.

Finally, there is Newton's method and its application to finding powers or nth roots of a number N (used in Chapter 2). Though it can be used effectively for other numerical evaluations, Newton's method does involve evaluating a derivation and can be a slow-converging process. Once it begins to close in on a root, however, it does so essentially doubling the number of decimal places to which we know the value of the root at each step. In this sense, it is a fast-converging method, once it gets close to the root.

In evaluating the complex roots of a function, we restrict ourselves to a method whose search pattern determines the roots only crudely. Their accurate evaluation is possible, but the necessary procedures and tracking methods are too complex for pocket calculator analysis. The approximation is usually sufficiently accurate for engineering analysis however.

The chapter closes with a discussion of the zeros of nth-order polynomials (we covered only first-, second-, third-, and fourth-order polynomials in Chapter 2). The search for the zeros of a polynomial is treated as a special case in this chapter because much is known about the roots of polynomials. Specifically, we know the following:

1. Polynomials of nth order have exactly n roots; thus we know precisely when all roots have been found.

2. When a zero is found, we can divide it out of the original polynomial to obtain a lower-degree polynomial for an easier evaluation of the remaining roots.

3. Most polynomials can be scaled to facilitate root evaluation.

4. The polynomial can occasionally be factored into its real linear and real quadratic factors—an important simplification in evaluating the roots of polynomials.

9-2 THE REAL ROOTS OF CONTINUOUS FUNCTIONS

To find the real roots of continuous functions mathematically, we must find a number x that, on substitution into a function $f(x)$, results in exactly zero. On a pocket calculator it is enough to find neighboring values of x that, when substituted into $f(x)$, provide nearly zero results of opposite signs. In this case we can approximate the zero of the function by using the midvalue of the interval.

In evaluating the zeros of mixed algebraic and transcendental functions the number of zeros is commonly found to be infinite. We must therefore identify the region in which the zero that we are interested in occurs. This can usually be done quickly by sketching the two functions or analytically determining by trial and error the neighborhood of our zero, from which we then begin the search for the root. For example, the function

$$f(x) = \frac{1}{\sin x} - \frac{1}{2}\left(\frac{x}{\pi} + 3\right)\sin x \tag{9-1}$$

has an infinite number of roots, as seen in Figure 9-1. The determination of which root to evaluate must be left to the analyst.

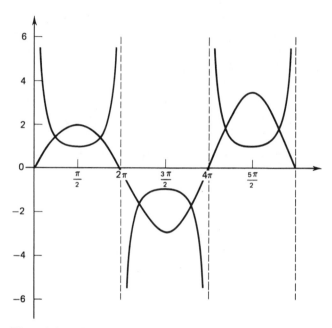

Figure 9-1   A function that has an infinite number of roots.

The simplest method of finding the real zeros of $f(x)$ is bisection. First, we identify an interval

$$x_1 \leqslant x \leqslant x_2$$

such that the product

$$(f(x_1))(f(x_2)) < 0 \tag{9-2}$$

That is, on one boundary of the interval $f(x)$ is positive and on the other boundary it is negative. Clearly, for the root to exist between the end values, this condition must hold true. This is seen in Figure 9-2. Once the interval is identified, another interval that satisfies the same property (equation 9-2) and is smaller is developed by way of the bisection method. That is, we evaluate the function at the midpoint of the interval and test it

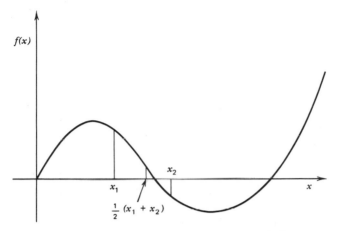

Figure 9-2   Searching for zeros using the bisection method.

to verify that the function is zero. If it is not (the usual case), the midpoint can be used as one end of a new interval, which will be smaller than the previous one. In particular, if the functions evaluated at both the midpoint and the original left end point of the initial interval are of the same sign, the zero lies to the right of the midpoint and the new interval is the previous right boundary and the midpoint. If, however, the function evaluated at the midpoint is of opposite sign to that evaluated at the left end point, the zero is to the left of the midpoint and the new interval is the midpoint and the previous left end point. All of this can be summarized as follows:

$$
\left(f\left(\frac{x_1 + x_2}{2} \right) \right) (f(x_1))
$$

$$
= \begin{cases} 0 & \frac{x_1 + x_2}{2} \text{ is a zero of } f(x) \\ >0 & \text{the zero is on the new interval } \left(\frac{x_1 + x_2}{2}, x_2 \right) \\ <0 & \text{the zero is on the new interval } \left(x_1, \frac{x_1 + x_2}{2} \right) \end{cases} \quad (9\text{-}3)
$$

Each iteration halves the length of the interval. Ideally, the initial interval should be identified so as to limit the required number of iterations. For example, an interval that is an integer will be reduced one-eighth by three iterations, one-sixteenth by four iterations, and one-thirty-second by five iterations. Interval reduction of one part in a thousand can be achieved with 10 iterations. The general formula is that the *interval size* can be reduced by a factor of $1/2^n$, where n is the number of iterations.

The bisection method raises the issue of how to end the iterative evaluation of the root. There are five common techniques. The first and most attractive approach is to specify the number of evaluations and then decide whether the bisection method is converging slowly enough to warrant finding another method. The second method is to test the absolute accuracy in x, that is, to determine whether the modulus of the differences between solutions of the midvalue is less than or equal to a small number. Another approach differs in relative accuracy; we test to see if a modulus of the difference in the successive values of x_n divided by x_{n-1} is less than or equal to ζ. Still other tests are intended to determine (1) whether $f(x_n)$ is less than or equal to some acceptable value and (2) whether the difference between successive values of $f(x)$ is less than or equal to some acceptable value.

These methods have all been used on digital computers to stop the iterative solutions of $f(x)=0$. In evaluating roots on the pocket calculator, however, tabulated values of x_n and $f(x_n)$ can be quickly computed, so that the convergence process becomes apparent to the analyst. In fact, the analyst generally stops iterating when the law of diminishing returns takes over and he sees little improvement in his evaluation of the roots for each iteration of the method.

About the only restriction on the bisection method is that in finding the zeros of functions with poles the bisection method locates the pole in a manner similar to that used to locate the zero (i.e., for functions where the approach to the pole from the right is positive and that from the left is negative).This problem will probably not be encountered, since the pocket calculator analyst will have at least sketched the function whose root he is trying to evaluate, thus knowing the characteristics of the function near the root.

9-3 FALSE-POSITION METHOD

The false-position method, sometimes called *regula falsi*, is based on the concept that, (a) when decreasing the interval in which the root is expected to be found and, (b) when the value of the function at one end of the interval is large compared with that at the other end, then the zero can be expected to be closer to the end where the function is small than to the end where the function is large. Interpolating between the values of the function at the end points of the interval, we can thus solve for the point at which this straight line passes through zero, using the equation

$$y(x)=f(x_1)+\frac{f(x_2)-f(x_1)}{x_2-x_1}(x-x_1) \qquad (9\text{-}4)$$

The line described by equation 9-4 has the zero at

$$x = \frac{x_1 f(x_2) - x_2 f(x_1)}{f(x_2) - f(x_1)} \tag{9-5}$$

Having found the new estimate of the location of the roots, we evaluate the function at this point and use it to replace the previous end point whose function value has the same sign as $f(x)$. At the same time we divide the function value of the other end point by 2.

Figure 9-3 shows the selection of the new estimate of the root, and Figure 9-4 illustrates the process by which the interval is halved and the approximating line is shaped to permit rapid convergence of the estimation of the zero.

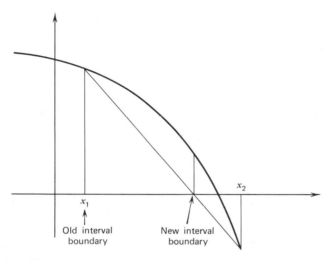

Figure 9-3   Searching for roots using the modified false position method.

9-4 NEWTON'S METHOD

In Newton's method the root is estimated and the tangent line of the function is computed at that point. Then the tangent line is projected until it intercepts the X axis to determine a second estimate of the root. Again, the derivative is evaluated and a tangent line formed to proceed to the third estimate of x. This process is sketched in Figure 9-5. The procedure is straightforward, but does involve the evaluation of derivatives. The tangent

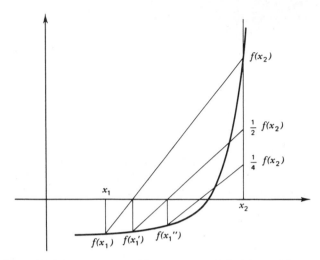

Figure 9-4   Interval halving to ensure rapid convergence of the false position method.

line generated in this manner is given by

$$y(x) = f(x_n) + f'(x_n)(x - x_n) \tag{9-6}$$

which, when $y(x) = 0$, gives the recursion formula for iterative estimates of the root:

$$x_{n+1} = x_n - \frac{f(x_n)}{f'(x_n)} \tag{9-7}$$

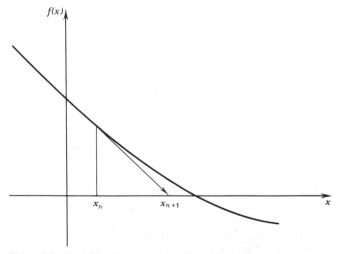

Figure 9-5   Searching for zeros using Newton's method.

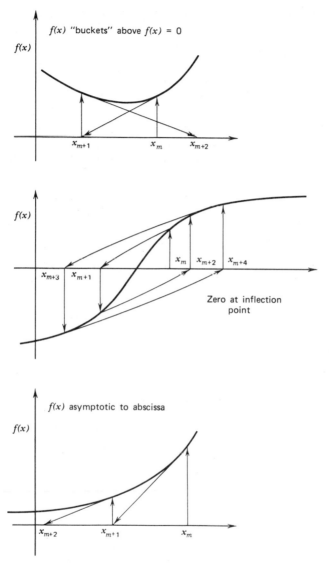

Figure 9-6 Problems that can be encountered using Newton's method.

Sketches of the well-known cases where Newton's method encounters difficulties appear in Figure 9-6, which emphasizes the analyst's need to sketch the function that he is trying to evaluate. Unless the local structure of the function is well understood, Newton's method should be avoided. The advantage to Newton's method is that, unlike the bisection method or the modified false-position method, once it begins to converge on the root, it tends to do so very quickly. In fact, at each step it almost doubles the number of accurate decimal places in the estimate of the root. Thus with an accuracy to 3 places at one step we can expect 6 places at the next step and 12 places thereafter. When it works, Newton's method is excellent.

9-5 COMPLEX ZEROS

To find complex zeros of analytic functions, we use the conventional complex variable notation—the independent variable is $z = x + iy$ and the dependent variable is $w = f(z) = f(x + iy)$, which is equal to $u(x,y) + iv(x,y)$. The condition

$$f(x + iy) = 0 \qquad (9\text{-}8)$$

is then equivalent to the two conditions

$$u(x,y) = 0 \qquad (9\text{-}9)$$

$$v(x,y) = 0 \qquad (9\text{-}10)$$

Since the equation $u(x,y) = 0$ defines curves in the complex plane and the equation $v(x,y) = 0$ defines another set of curves, it is only at the intersection of these two sets of curves that $w(z) = 0$. In this sense, the problem of finding the complex roots of $w = f(z)$ is equivalent to finding the intersection of the two curves $u = 0 = v$. Obviously, this is the problem of the simultaneous solution of two equations. Provided that the zero is not on the real axis and that $z = x + iy$ is zero, the conjugate $z_c = x - iv$ is also zero.

The bisection method can be extended from the problem of finding real zeros to that of finding complex zeros. Again, we first find the interval in which we can expect to find a zero and then refine the estimate of the zero by reducing the interval in which the zero is expected. In the bisection method, we first searched the region of the real axis where we expected roots to occur, not by evaluating the function numerically, but by determining the interval during which $f(x)$ changes sign. Then we narrowed that interval until we found the location of the root (as close as we wished). In a similar manner we break up the complex plane into regions where we can test, not the value of the complex function, but only its sign at suitably chosen points in the region and where we can determine the "quadrant-number" in which $w = f(z)$ falls. In general, if $w = f(z)$ is not zero (otherwise we would tabulate a zero at the grid point where it does equal zero), the quadrant numbers are defined as shown in Figure 9-7 and as they would be located in the complex plane in Figure 9-8. It is apparent that where the four quadrants meet we have a zero of $w = f(z)$. Once the grid is prepared and the quandrant numbers are written on the grid, the curves $u = 0$ and $v = 0$ can usually be quickly sketched by keeping in mind that the curves $u(x, y) = 0$ divide quadrants 1 and 2 and quadrants 3 and 4, whereas $v(x, y) = 0$ divide quadrants 1 and 4 and 2 and 3. Although this method is not very sophisticated, it is a convenient way of approximating roots in the complex plane on the pocket calculator. We therefore leave the more accurate evaluation of the complex roots to methods presented in other books. It is worth pointing out, however, that when a complex root is identified in this manner the region in the neighborhood of the root can be further subdivided to form a refined grid for more accurate root determination.

An example of the use of this method is finding the complex zeros of the function

$$w = az + b$$

The zeros of this complex function are easy to derive; the example is chosen for its pedantic value in illustrating the process of sketching the $u = 0 = v$ curves to find the region in which the complex zero will occur (which can then be used to make the next more refined search for the

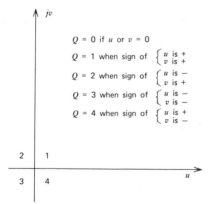

$Q = 0$ if u or $v = 0$

$Q = 1$ when sign of $\begin{cases} u \text{ is } + \\ v \text{ is } + \end{cases}$

$Q = 2$ when sign of $\begin{cases} u \text{ is } - \\ v \text{ is } + \end{cases}$

$Q = 3$ when sign of $\begin{cases} u \text{ is } - \\ v \text{ is } - \end{cases}$

$Q = 4$ when sign of $\begin{cases} u \text{ is } + \\ v \text{ is } - \end{cases}$

Figure 9-7 Quadrant number definitions.

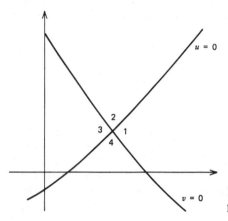

Figure 9-8 Quadrant numbers in the complex plane.

complex zero of w, etc.). First it is instructive to analytically find the zero of w. Substituting $(x + iy)$ for z we find

$$w = (ax + b) + i(ay) = u(x,y) + v(x,y)$$

Now

$$u = 0 = ax + b$$

$$\therefore x = -\frac{b}{a}$$

defines the $u = 0$ curve in the complex plane. Similarly

$$v = 0 = iay$$

or

$$iy = 0$$

defines the $v = 0$ curve in the complex plane. The intersection of these two curves is the zero of the function $w = 0$. This occurs at

$$x = -\frac{b}{a}$$

$$iy = 0$$

Now let us examine the use of the modified bisection method to sketch the $u = 0 = v$ curves for this function.

Table 9-1 shows the details of generating the quadrant numbers for 25 test locations in the complex plane. These locations are the *grid points* in a five-by-five test space:

$$x = -2, -1, 0, +1, +2$$

$$y = -1, 0, +1, +2, +3$$

The calculation of u and v and the analytical determination of the quadrant number using $\pm\tan(u/v)$ are straightforward on the scientific pocket calculator, particularly those with rectangular-to-polar conversion. The procedure for computing the quadrant numbers is shown in Table 9-2.

Table 9-1 Quadrant Numbers for the Function $\omega = az + b = (ax + b) + (iby) = u(x, y) + b(x, y)$ **When** $a = 1$ **and** $b = 1$

x	y	u	v	$\pm\tan^{-1}(v/u)$	Quadrant Number by Calculation — Quadrant Number	Quadrant Number by Inspection of Signs of u and v (sign u, sign v)
-2	-1	-1	-1	$-135°$	3	$(-,-)\to 3$
-1	-1	0	-1	$-90°$	0	$(0,-)\to 0$
0	-1	1	-1	$-45°$	4	$(+,-)\to 4$
$+1$	-1	2	-1	$-27°$	4	$(+,-)\to 4$
$+2$	-1	3	-1	$-18°$	4	$(+,-)\to 4$
-2	0	-1	0	$180°$	0	$(-,0)\to 0$
-1	0	0	0	Undefined	0	$(0,0)\to 0$
0	0	1	0	$0°$	0	$(+,0)\to 0$
$+1$	0	2	0	$0°$	0	$(+,0)\to 0$
$+2$	0	3	0	$0°$	0	$(+,0)\to 0$
-2	$+1$	-1	$+1$	$135°$	2	$(-,+)\to 2$
-1	$+1$	0	$+1$	$90°$	0	$(0,+)\to 0$
0	$+1$	1	$+1$	$45°$	1	$(+,+)\to 1$
$+1$	$+1$	2	$+1$	$27°$	1	$(+,+)\to 1$
$+2$	$+1$	3	$+1$	$18°$	1	$(+,+)\to 1$

Table 9-2 Procedure for Computing Quadrant Numbers

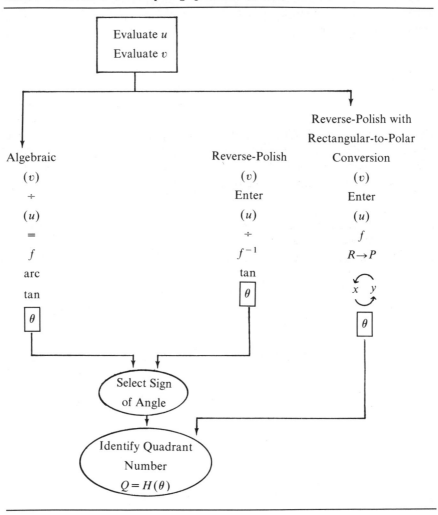

( )→data input.

☐ →output.

◯ →mental step done
  by operator.

Here $H(\theta)$ is given by

$$Q=1 \qquad \text{if } 0° < \theta < 90°$$

$$Q=2 \qquad \text{if } 90° < \theta < 180°$$

$$Q=3 \qquad \text{if } \begin{cases} 180° < \theta < 270° \\ -90° > \theta > -180° \end{cases}$$

$$Q=4 \qquad \text{if } \begin{cases} 270° < \theta < 360° \\ 0° > \theta > -90° \end{cases}$$

$$Q=0 \qquad \text{if } \theta = 0, 90, 180, 270, \text{ or } 0$$

Though computing the quadrant number requires only a few key strokes on the scientific calculator, and is a systematic analytical procedure, quadrant numbers are more quickly determined by inspecting the signs of u and v. Quadrant number determination by calculation is shown here more for the sake of completeness than utility.

Figure 9-9 shows the array of quadrant numbers located at their respective test points. Note that the line separating the quadrants 1 and 4, and 2 and 3, is the $v=0$ line, while the line separating the quadrants 1 and 2, and

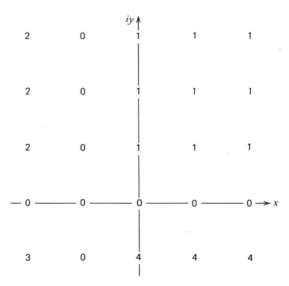

Figure 9-9 Quadrant numbers and root location for complex function $w = az + b$ where $a = 1$ and $b = 1$.

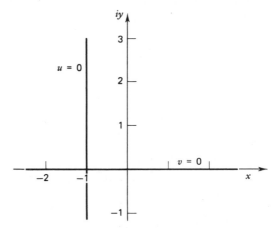

Figure 9-10 The lines $u=0$ and $v=0$ sketched from the matrix of quadrant numbers shown in Figure 9-9 for the complex function $w = az + b$ $(a=1, b=1)$.

3 and 4, is the $u=0$ line. These lines are sketched in Figure 9-10. While $w = az + b$ is a simple function whose zero is easy to determine, the procedure is identical for complex functions of more complicated forms. Hamming has worked a transcendental equation that illustrates this method very well. Figure 9-11 shows the array of quadrant numbers for the

```
                                                                    4

                                              4   4   4   3   3
                                  iy
  1   1   1   1   1    4   4   4   4   4   4   3   3   3   3   3

  1   1   1   1   1    4   4   4   4   4   3   3   3   3   3   3

  1   1   1   1   1    1   4   4   4   3   3   3   3   3   3   3

  2   2   1   1   1    1   4   4   3   3   3   3   2   2   2   2

  2   2   1   1   1    1   1   4   3   3   2   2   2   2   2   2

  2   2   0   1   1    1   1   4   4   2   2   2   2   2   2   2

  2   2   2   1   1    1   1   1   1   1   1   1   1   1   1   1

  2   2   2   2   1    1   1   1   1   1   1   1   1   1   1   1

 —0—0—0—0—0—0—0—0—0—0—0—0—0—0—0—0→
                                                                   x
```

Figure 9-11 Quadrant number array associated with $w = e^z - z^2$.

function

$$w = e^z - z^2$$

in the neighborhood of $x = iy = 0$. By inspection of the quadrant number field it is possible to sketch the $u = 0 = v$ curves, as seen in Figure 9-12.

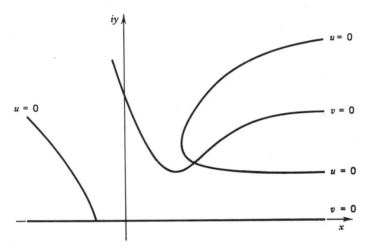

Figure 9-12   The curves $u = 0 = v$ sketched from the matrix of quadrant numbers shown in Figure 9-11 for the complex function $w = e^z - z^2$.

9-6. AN IMPROVED SEARCH METHOD

This unsophisticated approach to identifying the complex roots of a function has one major fault. The evaluation of the quadrants for each grid point requires a number of calculations to determine points that do not lie near the zeros of the complex function and thus give relatively little information about the location of the zeros. An alternative is to track the $u = 0$ curve and to identify, by marking the spot where this curve crosses the $v = 0$ curve, the region where the root will exist. The approach is to search a $u = 0$ curve in a counterclockwise direction in the area we are examining. The $u = 0$ curve will be indicated by a change from quadrant 1 to 2 (or 2 to 1) or from 3 to 4 (or 4 to 3). When we find the curve, we track it until we meet the $v = 0$ curve. The $v = 0$ curve will be indicated by the appearance of a new quadrant other than the two that we were using to track the $u = 0$ curve.

The procedure is straightforward and requires little practice to learn to conveniently track the $u = 0$ curve. In fact, the analyst learning to develop this method should take known u and v functions and practice tracking them until facility is developed in tracking unknown $u = 0$ curves.

9-7 PROBLEMS IN DETERMINING THE ZEROS OF POLYNOMIALS

In Chapter 2 we discussed methods for computing the roots of algebraic equations of up to the fourth order. The methods presented in this chapter permit the evaluation of the roots of higher-order polynomials. Polynomials are treated as special functions in determinining the zeros of functions because much is known about their zeros. The fundamental theorem of algebra guarantees that an nth-order polynomial will have exactly n roots. Once a zero is found, it can be "divided" out of the polynomial, thus reducing the order of the polynomial to a simpler form:

$$\frac{P_n(z)}{(z - z_1)} = P_{n-1}(z)$$

Also, all polynomials with real coefficients (those studied here) can be made up of linear and quadratic factors. If a polynomial is of odd order, at least one factor is linear and one root (at least) is real. Thus we must find the real root and divide it out of the polynomial. When the polynomial has been reduced to an even-order polynomial whose roots are either pairs of complex conjugate roots, pairs of real and equal roots, or pairs of real but unequal roots, we "scale" the polynomial so that the roots lie in a region that can be conveniently tested for the presence of additional real roots (they occur in pairs). When all real roots are identified, we merely need to find the complex roots for the remaining quadratic factors of the even-order polynomials.

Even when numerical methods are available for determining the roots of polynomials, we still encounter significant difficulty in the computing aspect of this task. To solve linear differential equations with constant coefficients, commonly the indicial equation is developed and solved for the characteristic roots of the system. These roots are then used in the assumed solution function to determine the solution of the differential equation. If the characteristic roots of a second-order differential equation are both real and equal, the solution should take the form

$$y = e^{-kx}(c_1 + c_2 x)$$

If, however, any numerical error enters into the pocket calculator evaluation of the roots, two real and slightly unequal roots of the form result

$$\text{root}_1 = -k + \epsilon$$

$$\text{root}_2 = -k - \epsilon$$

then the solution of the equation becomes

$$y = c_1 e^{-(k+\epsilon)x} + c_2 e^{-(k-\epsilon)x}$$

which is unlike the dynamics of a process with two real and equal roots. In this example, the purpose was to evaluate the characteristic roots of the differential equation, and even the smallest error affects the dynamics of the resultant solution to the differential equation. In another case we may be interested in the values of x and y that satisfy a system of simultaneous equations where small errors in the zeros result only in small errors in the evaluation of the simultaneous solution to the simultaneous equations.

All this serves to illustrate the following important point: the problem of finding the zeros of a function means different things to different people. It is necessary to define precisely what is required of the analyst when trying to evaluate the roots of an equation (to find the zeros of a function).

Hamming further brings out this important point by noting that theorems in mathematics do not necessarily apply to computing. In mathematics the notion of zeros of a function is a simple one, but not so in computing. Hence the answers to the question, "What is wanted of the zeros of a polynomial $P(x)$?" are usually as follows:

1. Those values of x_i that make $|P(x_i)|$ small should be as accurate as required.

2. At the values x_i the polynomial $P(x_i)$ should be as small as required.

3. The polynomial may be constructed as accurately as required from the zeros.

4. The zeros satisfy the auxiliary conditions for roots of polynomials (see Chapter 2).

What is required for one problem may not be required for another. Generally, it is thought that (1) is the answer, but actually (3) or (4) is really what is usually sought in applied analysis (as shown here in the differential equation example).

9-8 REFERENCES

For this chapter refer to Richard Hamming's *Numerical Methods for Scientists and Engineers* (McGraw-Hill, 1973), Chapters 4, 5, and 6.

CHAPTER 10

STATISTICS AND PROBABILITY

10-1 INTRODUCTION

The determination of the statistics of finite data populations is the topic of this chapter. For very large data populations probability is emphasized. Since the formulas used in statistical analysis are uncomplicated and for the most part directly implemented on the pocket calculator, we focus on statistical analysis more than on analysis on the pocket calculator. The objective is to provide the pocket calculator owner with a classical basis of scientific statistical analysis. However, "tricks" of the pocket calculator trade will be mentioned whenever applicable. Also, when there are a number of alternative ways to compute a statistic, emphasis is placed on the formulas that are most easily evaluated on the pocket calculator.

First we discuss the numerical evaluation of the statistics that characterize data populations. Emphasis is on measures of central tendency, measures of dispersion, data distributions, shapes of data distributions, and the elements of probability. Then we proceed to the concepts of sampling and testing.

Throughout the chapter the focus is on the statistical analysis of groups of data and their functional interpretation in engineering and scientific analysis. As in Chapter 6, we emphasize the following:

1. Understanding statistical analysis on the pocket calculator.
2. Providing useful formulas and tables of data for statistical analysis.

10-2 FREQUENCY DISTRIBUTIONS

A number of definitions and concepts are prerequisite to a discussion of the formulas for computing the statistics of data populations. The definitions presented here are working definitions and *are not* presented in abstract mathematical notation. We wish to impart a working knowledge of statistical analysis on the pocket calculator, rather than a theoretical knowledge of the field of statistics and probability.

A basic data population is simply a collection of statistics called the raw data. Raw data are data that lack organization. Arrays and frequency distribution are ways of organizing the data, so that the statistics of a collection of data can be determined.

Arrays are arrangements of raw data in ascending or descending order; that is, the data are tabulated starting with the largest number and proceeding to the smallest, and vice versa. We say that the range of an array is the difference between the largest and smallest numbers in the array.

When large sets of data are stratified into categories and the number of elements in the data set belong to each category or class, we form a data distribution. This is done by generating a table of data by category or class, together with the class frequencies, that is, the number of elements of the set of all data belonging to each class. Such a tabular array is called a frequency distribution or frequency table. An example appears in Table 10-1. Here the classes or categories are the intervals of height. Data arranged in a frequency distribution are often also called grouped data. The term "class mark" refers to the midpoint of a class interval.

In general, frequency distributions are developed by first determining the range of the raw data, dividing the range into a convenient number of class samples of the same size, and then determining the number of observations that fall into each class interval (this, by definition, is called the class frequency). Once the frequency distribution is known, *histograms* can be developed to visualize the frequency distribution. Histograms are simply a plot of the frequency against the range of raw data. An example of a histogram for the frequency distribution in Table 10-1 is shown in Figure 10-1.

Table 10-1   Heights of 100 Male Students in a University

Height (in.)	Number of Students	Cumulative Number of Students
60–63	4	4
63–66	18	22
66–69	41	63
69–72	28	91
72–75	9	100

The relative frequency of a class is its frequency divided by the total frequency of all classes. It is multiplied by 100 to obtain a percentage. Plots of relative frequency over the range of the data are called a percentage distribution or relative frequency distribution.

The cumulative frequency distribution is simply defined as the total frequency of all data less the upper class or category boundary of a given class interval. The third column of Table 10-1 shows the cumulative frequency distribution of the height of the 100 students. A cumulative frequency distribution can be plotted over the range of data as shown in Figure 10-2.

Relative cumulative frequency distributions are defined like the frequency distributions, and so is their percentage.

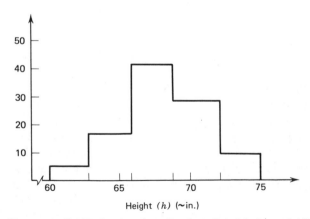

Height (h) $(\sim$in.$)$

Figure 10-1   Frequency distribution (number of students in height interval Δh).

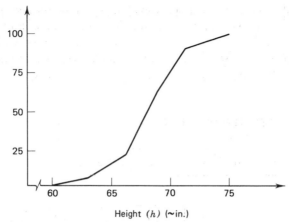

Height (h) $(\sim$in.$)$

Figure 10-2 Cumulative frequency distribution (cumulative number of students with heights less than or equal to h).

Were the data to increase without bound, the histogram's frequency distribution and cumulative frequency distribution would be expected to be developed with increasingly finer quantitization until smooth curves are obtained. The analysis of probability using continuous functions is the field of probability analysis. Here we concentrate on the statistics associated with finite sized data sets. We examine probability as well, but our emphasis is on the statistical analysis of small sized data sets that can be reasonably analyzed on the pocket calculator.

10-3 MEASURES OF CENTRAL TENDENCY

The mean, median, mode, and other *measures of central tendency* are the *statistics* of data distributions. Numbers that inform us about the centroid of the distributions are called measures of central tendency. There are not a few such measures, all called averages. The most common of them have already been discussed, but we mention them again in the context of this chapter. They are the arithmetic mean, the geometric mean, and the harmonic mean. Two other measures of central tendency that are important in statistics are the median and the mode. The arithmetic mean is defined by the relationship

$$\bar{X} = \frac{\sum_{j=1}^{N} X_j}{N} = \frac{\sum X}{N} \tag{10-1}$$

The arithmetic mean can be quickly evaluated, as mentioned in Chapter 3, either recursively or by simply summing all of the samples and dividing by the total number of samples summed. For calculators with the $\boxed{\Sigma}$ function (shown as $\boxed{M+}$ on many calculators) the inclination is to perform the arithmetic mean with the key stroke sequence

ALGEBRAIC KEY STROKE SEQUENCE

$$(X_1)\boxed{f}\boxed{M+}(X_2)\boxed{f}\boxed{M+}(X_3)\boxed{f}\boxed{M+}\cdots$$

$$(X_n)\boxed{f}\boxed{M+}\boxed{f}\boxed{M\to x}\boxed{\div}(N)\boxed{=}\boxed{\bar{X}}$$

for a total of $(2N+6)$ function key strokes and $(5N+5)$ data entry key strokes (assuming 5 digits for every data entry), totaling $(7N+8)$ key strokes. With the straight arithmetic sum, however, the key stroke sequence is

$$(X_1)\boxed{+}(X_2)\boxed{+}(X_3)\boxed{+}\cdots\boxed{+}(X_n)\boxed{\div}(N)\boxed{=}$$

and involves only $(6N+6)$ key strokes, thus saving 15 to 20% in key strokes. The benefits of the Σ or $M+$ functions are only accrued when the sum

$$\Sigma f(X) = \sum_{j=1}^{N} f(X_i), \quad (f(X_i) \neq X_i)$$

is computed. Then the intermediate calculation for evaluating the $f(X_i)$ can be conducted without disturbing their accumulation.

If certain numbers occur more than once, in particular with frequencies $f_1, f_2, \ldots, f_n$, then the arithmetic mean is defined by

$$\bar{X} = \sum_{j=1}^{K} f_j X_j / \sum_{j=1}^{K} f_j = \frac{\Sigma fX}{\Sigma f} = \frac{\Sigma fX}{N} \tag{10-2}$$

When the weighting factors are associated with certain numbers to emphasize or change their importance in the distribution, the weighted arithmetic mean is defined as

$$\overline{X} = \frac{\sum WX}{\sum W} \qquad (10\text{-}3)$$

This is called the weighted arithmetic mean.

The median of an array of numbers (i.e., numbers arranged in order of increasing or decreasing size) is defined to be the middle value of the array. If the array has an odd number of elements, there is a single middle value. If the array has an even number of elements, there are two middle values, in which case the median is defined as the average of the two middle values.

From a histogram viewpoint, the median is that value of the data range which exactly divides the histogram into two equal parts. The mode of a distribution is that value which occurs most frequently.

In general, distributions can have more than one mode, but only one mean and one median.

For unimodal distributions that are only slightly asymmetrical (skewed) the mean, median, and mode are approximately related by the relation

$$\text{mean} - \text{mode} = 3(\text{mean} - \text{median}) \qquad (10\text{-}4)$$

The distinction between the mean, median, and mode is shown in Figure 10-3.

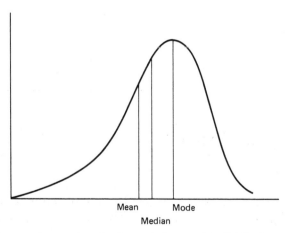

Figure 10-3 Mean, median, and mode of a typical symmetric distribution.

Another frequent measure of central tendency is the harmonic mean defined by the relation

$$H = \frac{1}{\frac{1}{N}\sum 1/X} = \frac{N}{\sum 1/X} \tag{10-5}$$

From time to time the mean of the logarithm of a set of numbers x is to be computed; then the geometric mean of the numbers is given by

$$G = (X_1, X_2 \cdots X_N)^{1/N} \tag{10-6}$$

The recursion formulas for computing the geometric and harmonic mean are presented in Chapter 3. Computing the harmonic mean using the Σ (or $M+$) function on the scientific pocket calculator is an excellent example of how this function saves key strokes. The key stroke sequence is

ALGEBRAIC KEY STROKE SEQUENCE

$$(X_1)\boxed{f}\boxed{1/x}\boxed{f}\boxed{\Sigma}(X_2)\boxed{f}\boxed{1/x}\boxed{f}\boxed{\Sigma}\cdots$$

$$(X_N)\boxed{f}\boxed{1/x}\boxed{f}\boxed{\Sigma}\boxed{f}\boxed{M\rightarrow x}(N)\boxed{f}\boxed{1/x}\boxed{=}\boxed{H}$$

as opposed to

$$(X_1)\boxed{f}\boxed{1/x}\boxed{f}\boxed{X\rightarrow M}(X_2)\boxed{f}\boxed{1/x}\boxed{+}\boxed{f}\boxed{M\rightarrow X}\boxed{=}\boxed{f}\boxed{X\rightarrow M}(X_3)\cdots$$

$$(X_N)\boxed{f}\boxed{1/x}\boxed{+}\boxed{f}\boxed{M\rightarrow X}\boxed{=}\boxed{\div}(N)\boxed{f}\boxed{1/x}\boxed{=}\boxed{H}.$$

The number of key strokes with the $\boxed{M+}$ operator is $(6N+11)$ and $(8N+16)$ without. As much as 33% fewer key strokes are required when the $\boxed{M+}$ function is available on the keyboard.

In reverse-polish the sequence of key strokes is

REVERSE-POLISH KEY STROKE SEQUENCE

$(X_1)\boxed{1/x}\boxed{\uparrow}(X_2)\boxed{1/x}\boxed{+}(X_3)\boxed{1/x}\boxed{+}\cdots$

$(X_n)\boxed{1/x}\boxed{+}(N)\boxed{\div}\boxed{1/x}\boxed{H}$

for a total of $(7N+7)$ key strokes.

As mentioned in Chapter 3, the relation between the harmonic, arithmetic, and geometric means is

$$H \leqslant G \leqslant \overline{X} \tag{10-7}$$

when all of the numbers used for calculating these means are identical.

Another mean, used extensively in this chapter, is the root mean square. It is a set of numbers defined by the relation

$$\mathrm{RMS} = \left(\frac{\sum X^2}{N}\right)^{1/2} \tag{10-8}$$

The evaluation of the root mean square on many scientific calculators is simplified somewhat by the use of the $M + x^2$ function where the contents of the display register (X) are squared and added to whatever is in the memory register.

10-4 MEASURES OF DISPERSION

Dispersion is defined to be the distribution or spread of the data about the average value. It is also frequently called the variation of the data. The number of measures of dispersion or variation of data about the mean is almost as large as the numbers of great statisticians. Here, we are concerned with those that can be used in practical analysis, such as the range, mean deviation, and standard deviation of the data.

The range of the data was defined earlier. The *mean deviation* or average deviation of a set of numbers is defined by the relations

$$\text{mean deviation} = MD = \frac{\sum\limits_{j=1}^{N} |X_j - \bar{X}|}{N} = \frac{\sum |X - \bar{X}|}{N} = \overline{|X - \bar{X}|} = \text{avg}(|X - \bar{X}|)$$

$$(10\text{-}9)$$

The *standard deviation* is denoted by s and is defined by the relations

$$s = \left(\frac{\sum\limits_{j=1}^{N} (X_j - \bar{X})^2}{N-1} \right)^{1/2} = \left(\frac{\sum (X - \bar{X})^2}{N-1} \right)^{1/2} = \sqrt{\frac{\sum X^2}{N-1}} = \sqrt{\overline{(X - \bar{X})^2}}$$

$$(10\text{-}10)$$

Occasionally the standard deviation is written in the "biased-estimate" form:

$$s \cong \left[\frac{\sum (X - \bar{X})^2}{N} \right]^{1/2}$$

$$(10\text{-}11)$$

for $N > 30$.

Finally the variance of a set of data is defined as the square of the standard deviation. A better form for computing the standard deviation from the pocket calculator viewpoint is

$$s = \begin{cases} \left[\dfrac{\sum X^2}{N} - \left(\dfrac{\sum X}{N} \right)^2 \right]^{1/2} = \left[\overline{X^2} - \bar{X}^2 \right]^{1/2} \\[3em] \left[\dfrac{\sum fX^2}{N} - \left(\dfrac{\sum fX}{N} \right)^2 \right]^{1/2} \end{cases}$$

$$(10\text{-}12)$$

The standard deviation has a number of useful properties for computational analysis. Thus for normal distributions (we discuss them later in this chapter) 68.27% of all the cases are included between the mean $\pm s$, that is, 68.27% of all cases reside within one standard deviation of the mean.

Similarly, 95.45% of all the cases lie within $2s$ of the mean and 99.73% of all the cases lie within $3s$ of the mean.

If two distributions have total frequencies N_1, N_2, their variations s_1^2 and s_2^2 can be combined according to the relation

$$s^2 = \frac{N_1 s_1^2 + N_2 s_2^2}{N_1 + N_2} \tag{10-13}$$

when both distributions have the same mean (not an infrequent case). The generalization of this formula to n distributions is straightforward when it is noted that the variance of n distributions with the same mean is simply the weighted arithmetic mean of the individual variances, where the weighting factor is the frequency of each distribution.

As with absolute and relative error, we can quote absolute and relative dispersions. A measure of *relative dispersion* is given by

$$\text{relative dispersion} = \frac{\text{absolute dispersion}}{\text{average}} \tag{10-14}$$

If we use the measure of absolute dispersion to be the standard deviation and the average to be the mean, we can define the *coefficient of variation* to be

$$V = \frac{s}{\overline{X}} \tag{10-15}$$

Clearly, the coefficient of variation is useless for zero-centered symmetric distributions.

10-5 MEASURES OF DISTRIBUTION SHAPE

Before going into measures of skewness and kurtosis, we must define the moments of a distribution. The rth moment of a distribution consisting of n values is given by

$$\overline{X^r} = \frac{\sum\limits_{j=1}^{N} X_j^r}{N} = \frac{\sum X^r}{N} \tag{10-16}$$

The first moment where $r = 1$ is the arithmetic mean. For a nonzero mean we can further define the rth moment about the mean as

$$m_r = \frac{\sum\limits_{j=1}^{N} \left(X_j - \overline{X} \right)^r}{N} = \frac{\sum (X - \overline{X})^r}{N} = \overline{(X - \overline{X})^r} \tag{10-17}$$

Similarly, the rth moment about any origin A is defined by

$$m'_r = \frac{\sum_{j=1}^{N}(X_j - A)^r}{N} = \frac{\sum(X-A)^r}{N} = \frac{\sum d^r}{N} = \overline{(X-A)^r} \quad (10\text{-}18)$$

The moments for group data can be defined in a similar manner as

$$\overline{X^r} = \frac{\sum_{j=1}^{N} f_j X_j^r}{N} = \frac{\sum f X^r}{N} \quad (10\text{-}19)$$

$$m_r = \frac{\sum f(X - \overline{X})^r}{N} \quad (10\text{-}20)$$

$$m'_r \frac{\sum f(X - A)^r}{N} \quad (10\text{-}21)$$

Again note that these formulas are in forms immediately useful for evaluation on the pocket calculator without the need to rewrite them.

The degree to which a distribution is asymmetric is specified by the skewness of the distribution. That is, if the distribution has a longer tail to the right of the distribution centroid, the distribution is said to be skewed to the right and to have a positive skewness. Conversely, if the distribution has a longer tail to the left of its centroid, the distribution is said to be skewed to the left and to have a negative skewness. The mean tends to lie on the same side of the mode as the longer tail for skewed distributions. A measure of symmetry due to Pearson is therefore given by the difference between the mean and the mode. This difference is then divided by the standard deviation (to make the measure dimensionless) to form a measure of the skewness. Thus Pearson's measure of skewness is

$$\text{skewness} = \frac{\text{mean} - \text{mode}}{\text{standard deviation}} = \frac{\overline{X} - \text{mode}}{s} \quad (10\text{-}22)$$

Kurtosis is a measure of the degree to which the distribution is peaked. A common measure of kurtosis is

$$a_4 = \frac{m_4}{s^4} = \frac{m_4}{m_2^2} \quad (10\text{-}23)$$

That is, the fourth moment is divided by the fourth power of the standard deviation. For Gaussian distributions, the measure of kurtosis is 3. Kurtosis is then sometimes defined by the relationship

$$a_4 - 3 = \text{kurtosis}$$

The moment of kurtosis is also referred to as the *coefficient of excess*.

10-6 PROBABILITY

Now that we have developed the basic definitions in statistics, we can examine the elements of probability, which we later relate to statistics through concepts in sampling. We will then be ready to discuss information theory, decision theory, the elements of nonparametric statistics, and concepts of correlation. For our work here we define probability as follows. If an event can happen in h ways out of n equally likely ways, the probability of occurrence of this event is given by

$$p = \Pr\{E\} = \frac{h}{n} \tag{10-24}$$

we also say that this is the probability of success of the event; similarly $q = 1 - p$ is equal to the probability of failure of the event. Clearly, $p + q$ must equal 1.

If E_1 and E_2 are dependent events, that is, the probability that E_2 occurs given that E_1 has occurred is nonzero, we say that the probability E_2 will occur given E_1 is the conditional probability of E_2 given E_1 and is written

$$\Pr\{E_2|E_1\} = \frac{\Pr\{E_2 \text{ and } E_1\}}{\Pr(E_1)} \tag{10-25}$$

Now, the *multiplication law* or the law of compound probabilities (i.e., the probability that both E_1 and E_2 occur simultaneously) is given by

$$\Pr\{E_1 E_2\} = \Pr\{E_1\}\Pr\{E_2|E_1\} \tag{10-26}$$

and the *addition law* of probability (i.e., the probability that either E_1 or E_2 occurs) is given by

$$\Pr\{E_1 + E_2\} = \Pr\{E_1\} + \Pr\{E_2\} - \Pr\{E_1 E_2\} \tag{10-27}$$

If the events are mutually exclusive, then

$$\Pr\{E_1 + E_2\} = \Pr\{E_1\} + \Pr\{E_2\} \tag{10-28}$$

since $\Pr(E_1 E_2) = 0$. If the events are statistically independent, then

$$\Pr\{E_1 E_2\} = \Pr\{E_1\} \Pr\{E_2\} \tag{10-29}$$

since $\Pr(E_2 | E_1) = \Pr(E_2)$.

Clearly, for events that occur discretely, such as rolling dice and flipping coins, we can form a discrete probability distribution by tabulating the event and the probability that it will occur. An example appears in Table 10-2.

Table 10-2   Sum of Points on a Single Throw of Two Dice

Sum of Points $= X$	Pr(X)	Cum Pr(X)
2	1/36	1/36
3	2/36	3/36
4	3/36	6/36
5	4/36	10/36
6	5/36	15/36
7	6/36	21/36
8	5/36	26/36
9	4/36	30/36
10	3/36	33/36
11	2/36	35/36
12	1/36	1

Also, discrete probability distributions can be plotted like histograms. Cumulative probability distributions can be developed in the same way as our cumulative relative frequency distributions. Then, as the number of probable events approaches infinity, the probability distribution functions become increasingly dense until we can form continuous probability distribution as the limit to which the discrete probability distribution approaches as the quantitization in the process goes to zero.

A concept commonly used in probability is that of expectation. Expectation is defined as follows. If X denotes a discrete random variable that can take on values $X_1, X_2, \ldots, X_k$ with associated probabilities $p_1, p_2, \ldots, p_k$, where $p_1 + p_2 + \cdots + p_k = 1$, the expectation of X is defined as

$$E(X) = \sum_{j=1}^{k} p_j X_j = \sum pX \qquad (10\text{-}30)$$

The extension of equation 10-30 to continuous distribution is obvious. In this case, we would define $E(x) = \mu(\text{mu})$, the mean of the population; m (the mean of the sample) is an estimate of the true $E(x)$.

It should be clear from this discussion a very large random sample of size N from a population would result in a sample mean that is very near the population mean, whereas a sample mean based on a small sample would not be likely to be very near the population mean.

10-7 PROBABILITY DISTRIBUTIONS

Of the many distributions used in probability and statistical analysis three are frequently encountered in engineering and scientific work: the binomial, the Gaussian, and the Poisson.

The binomial distribution is defined by

$$p(X) = {}_N C_X p^X q^{N-X} = \frac{N!}{X!(N-X)!} p^X q^{N-X} \qquad (10\text{-}31)$$

where p is the probability of success, that is, the probability that an event will happen in any single trial; q is the probability that it will fail to happen in any single trial, usually called the probability of failure and equal to $1 - p$; and $p(X)$ is the probability that the event will happen exactly X times in N trials. Here X is defined only on the integers, that is, $X = 0, 1, 2, \ldots, N$. By definition this is a discrete probability distribution. Its name reflects the fact that as X takes on integer values from 1 through N, the corresponding probabilities are given by the terms in the binomial expansion

$$(p + q)^N = q^N + {}_N C_1 pq^{N-1} + {}_N C_2 p^2 q^{N-2} + \cdots + p^N \qquad (10\text{-}32)$$

The statistics associated with the binomial distribution are developed in many books and will not be repeated here. Four are usually used in practical analysis, the mean, defined by

$$\mu = Np \tag{10-33}$$

the variance defined by

$$\sigma^2 = Npq \tag{10-34}$$

the coefficient of skewness defined by

$$a_3 = \frac{q - p}{(Npq)^{1/2}} \tag{10-35}$$

and the measure of kurtosis (or coefficient of excess) given by

$$a_4 = 3 + \frac{1 - 6pq}{Npq} \tag{10-36}$$

As an example of the application of the binomial distribution and its statistics, let us consider 100 flips of an unbiased coin. The probability is one-half that the coin will be heads and one-half that it will be tails. The mean thus is

$$\mu = Np$$

$$\mu = 100 \times \tfrac{1}{2} = 50$$

The standard deviation is given by the square root of the variance:

$$\sigma = \sqrt{Npq}$$

$$\sigma = \sqrt{100 \times \tfrac{1}{2} \times \tfrac{1}{2}} = \sqrt{25} = 5$$

Skewness is zero, since $p = q$, and the measure of kurtosis equals

$$\alpha_4 = 3 + \frac{1 - 6pq}{Npq}$$

$$\alpha_4 = 3 + \frac{(1 - \frac{3}{2})}{25}$$

$$\alpha_4 = 3 - \frac{1}{50}$$

$$\alpha_4 \approx 3$$

We see that the expected number of heads in 100 flips of a coin is 50. The standard deviation for 100 trials is 5. We would not expect the distribution of heads and tails to be skewed (if the coin is unbiased), but would expect it to be approximately Gaussian (the kurtosis of a Gaussian distribution equals 3).

The Gaussian distribution is defined by the equation

$$Y = \frac{1}{\sigma\sqrt{2\pi}} e^{-\frac{(x - \mu)^2}{2\sigma^2}} \qquad (10\text{-}37)$$

where

$$\mu = \text{the mean}$$

$$\sigma = \text{the standard deviation}$$

The Gaussian distribution has the following characteristics:

1. The area bounded by the distribution and the X axis is identically equal to 1.

2. The area bounded by the distribution and the X axis in the interval between $x = a$ and $x = b$ where $a < b$ is identically equal to the probability that X lies between a and b. The Gaussian distribution is shown in Figure 10-4.

The parameters of the Gaussian distribution are the mean and variance, defined by

$$\text{mean} = \mu$$

$$\text{variance} = \sigma^2$$

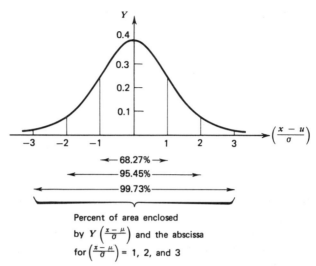

Figure 10-4   Gaussian distribution.

The skewness of the Gaussian distribution by inspection is zero. The measure of kurtosis for the Gaussian distribution is given by

$$a_4 = 3$$

As mentioned before, the binomial and normal distributions are related. If for a binomial random variable X, N, and P are not zero nor near zero and N is large, the binomial distribution can be approximated with the Gaussian distribution, since $(X - Np)/\sqrt{Npq}$ is approximately normal with 0 mean and unit (1) variance. It turns out that the Gaussian distribution is the limiting form of the binomial distribution as N approaches infinity. In practical work the Gaussian distribution is a reasonable approximation of the binomial distribution if both Np and Nq are greater than 5.

Another probability distribution encountered in practical probability work is the Poisson distribution, defined by

$$p(X) = \frac{\lambda^X e^{-\lambda}}{X!} \qquad (10\text{-}38)$$

Here λ is a constant of the distribution.

The statistics of the Poisson distribution are all given in terms of the single parameter λ. The mean is given by

$$\mu = \lambda \qquad (10\text{-}39)$$

The variance is given by

$$\sigma^2 = \lambda \qquad (10\text{-}40)$$

which is identically equal to the mean (an interesting curiosity of the Poisson distribution, which we discuss later). The coefficient of skewness of the Poisson distribution is

$$a_3 = \frac{1}{\sqrt{\lambda}} \qquad (10\text{-}41)$$

and the measure of kurtosis is

$$a_4 = 3 + \tfrac{1}{\lambda} \qquad (10\text{-}42)$$

A number of things become apparent by examining the Poisson distribution. First, it is discrete and its basic properties are not a function of the number of trials being considered. Second, the larger the mean (λ), the less the skewness and the more the distribution tends toward the Gaussian distribution; accordingly the coefficient of excess approaches 3. From our observations of skewness and kurtosis we might expect that the Gaussian distribution is the limit to which the Poisson distribution approaches as λ approaches infinity; this, in fact, turns out to be the case.

As might be expected, the binomial and Poisson distributions are related, both being discrete and both approaching the Gaussian distribution as a limit for large numbers samples (in the case of the binomial distribution) or a large value of the mean (in the case of the Poisson distribution). Note that in the binomial distribution if N is large and the probability p of occurrence of an event is close to zero, $q = (1 - p)$ is close to 1 and we say that the event is a rare event. In these situations the binomial distribution can be closely approximated by the Poisson distribution by letting $\lambda = Np$. Comparison of the mean, variance, skewness, and coefficient of excess when $\lambda = Np$, $p = 0$, and $q \cong 1$ shows that the binomial distribution properties (e.g., mean, variance) are approximately equal to those of the Poisson distribution. Extension of the Poisson distribution to its association with the Gaussian distribution follows through its association with the binomial distribution. After a little algebra, it can be shown that the Poisson distribution can be approximated by the Gaussian distribution, since

$$\frac{(X - \lambda)}{\sqrt{\lambda}}$$

is approximately normally distributed with zero mean and unit (1) variance. It is common in statistical analysis to use these distribution functions to model the distribution of populations being studied. The approach is to determine the mean and standard deviation of the sample of the population to estimate the mean and standard deviation of the population. The modeling process is then tested for goodness of fit, using a number of approaches. One that we discuss later in this chapter is the chi-square test. Clearly, modeling a sample of a population with the binomial or Poisson distribution solely amounts to determining the mean value $(\overline{X})$ of the sample distribution. For modeling the binomial distribution we compute

$$p = \frac{\overline{X}}{N}$$

Then

$$q = 1 - p$$

A crude test is to determine if

$$Nqp \approx s^2$$

where s is the standard deviation of the sampled population. If the calculation shows that

$$p \approx 0$$

then the Poisson distribution may better fit the data.

10-8 SAMPLING

The study of sampling deals with the determination of statistics (estimates of distribution parameters) associated with a sample drawn from a population and the population parameters themselves. We are concerned with sampling insofar as it is related to estimation, testing, and statistical inference. First we wish to define a sample, in particular a random sample. Generally it is not sufficient to sample a population in a systematic way. Since the objective is to develop the statistics of the sample and infer something about the parameters of the distribution, the sample that is representative of the population must be chosen. There are a number of methods for sampling a population, including stratified and random sampling. One universally accepted way of sampling a population so that the characteristics of the population are represented in the sample is *random sampling*. Random sampling is a process in which each member of the population has an equal chance of being included in the sample. Typically a number is assigned to each member of the population, and the numbers are then scrambled, so that each sample of the population has the same chance of being selected as any other sample.

There are two concepts of sampling: with and without replacement. The concepts are straightforward. Sampling with replacement allows each member of the population to be chosen more than once, while sampling without replacement does not. The analysis is useful because a finite sample with replacement can theoretically be considered infinite. Sampling without replacement will result in a statistic that takes into account the size of the sample compared with the total size of the population.

For any given sample of size N a value of a given statistic can be computed for that group and, if another sample of the same size is selected, in general a different value of the statistic is computed. This process can continue (with or without replacement) until all statistic values are in a population different. Clearly, if a population is of finite size and is sampled with replacement, an infinite number of values can be determined for each statistic. By organizing these values of the statistic, we obtain a distribution of the statistic itself, which is called its sample distribution. Clearly, then, there are sampling distributions of the mean, sampling distributions of the standard deviation, sampling distributions of the variance, sampling distributions of the measure of kurtosis, and sampling distributions of the coefficient of skewness. Of all these possible statistics we focus on those that help us in testing, estimating, and statistical inference only.

Suppose that we have a population of finite size, say N_p. Also suppose that a sample of size N is drawn without replacement. Then if we denote the mean of the sample distribution of the mean by

$$\mu_{\bar{X}}$$

and the standard deviation of the sample distribution of the mean as

$$\sigma_{\bar{X}}$$

and, further, if we define the population mean and standard deviation by μ and σ, respectively, we can write

$$\mu_{\bar{X}} = \mu$$

$$\sigma_{\bar{X}} = \frac{\sigma}{\sqrt{N}} \left(\frac{N_p - N}{N_p - 1} \right)^{1/2}$$

(10-43)

If the population is infinite or if it is finite but sampled with replacement, the population mean and standard deviations are related to the sample mean and standard deviations according to

$$\mu_{\bar{X}} = \mu$$

$$\sigma_{\bar{X}} = \frac{\sigma}{\sqrt{N}}$$

(10-44)

For N greater than 30, the sample distribution of the mean is approximately Gaussian with mean

$$\mu_{\text{Gaussian}} \cong \mu_{\bar{X}}$$

and standard deviation

$$\sigma_{\text{Gaussian}} \cong \sigma_{\bar{X}}$$

It is noted that the distribution of the mean of a population sample is independent of the distribution population.

Suppose now that we draw a sample from each of two populations. The number of samples drawn from the first population is N_1 and that from the second population is N_2. Now let us compute a statistic s_1. We find that each statistic has a sampling distribution, whose mean and standard deviation we denote by μ_{s_1} and σ_{s_1}. A similar situation holds true for the second population; that is, it has a mean and standard deviation given by μ_{s_2} and σ_{s_2}. If we now consider all possible combinations of these samples from the two populations, we can obtain a distribution of differences, that is, $S_1 - S_2$, which is called the sampling distribution of the differences of the statistics. The mean and standard deviation of this distribution are defined by equation 10-45:

$$\mu_{S_1 - S_2} = \mu_{S_1} - \mu_{S_2}$$

$$\sigma_{S_1 - S_2} = \left(\sigma_{S_1}^2 + \sigma_{S_2}^2 \right)^{1/2}$$

(10-45)

If S_1 and S_2 are the sample means from two populations, the sampling distribution of the differences of the means is given for infinite populations with mean and standard deviation μ_1, σ_1 and μ_2, σ_2:

$$\mu_{\bar{X}_1 - \bar{X}_2} = \mu_{\bar{X}_1} - \mu_{\bar{X}_2} = \mu_1 - \mu_2$$

$$\sigma_{\bar{X}_1 - \bar{X}_2} = \left(\sigma_{\bar{X}_1}^2 + \sigma_{\bar{X}_2}^2 \right)^{1/2} = \left(\frac{\sigma_1^2}{N_1} + \frac{\sigma_2^2}{N_2} \right)^{1/2}$$

(10-46)

These results hold for finite populations if sampling is with replacement.

The standard deviation of a population can also be computed from a sample of the population. It, too, has a distribution with mean and standard deviations

$$\mu_s = \sigma \cong s \left(\frac{N}{N-1} \right)^{1/2}$$

$$\sigma_s = \left(\frac{\mu_4 - \mu_2^2}{4 N \mu_2} \right)$$

(10-47)

For populations that are normally distributed this reduces to ($\mu_2 = \sigma^2$ and $\mu_4 = 3\sigma^4$):

$$\mu_s = \sigma$$

$$\sigma_s = \frac{\sigma}{\sqrt{2N}}$$

(10-48)

10-9 STATISTICAL ESTIMATION

We have just seen how statistical information can be computed from the data of the sampled population. One of the key problems in statistical inference is that of estimation of the population parameters (mean, variance, kurtosis, etc.) from the corresponding sample statistics. Before proceeding to the concept of confidence interval estimates for population parameters, the key issue in this section, we must clear up the issue of biased versus unbiased estimation. In computing a mean value one merely sums the numbers of the sampled values and divides by the total number. In computing the standard deviation, however, it is important to recognize that it takes at least two points to compute a variance; hence the mean of the variance of a distribution must be divided by $N-1$, not N. In this sense, then, we have an unbiased estimator. As N becomes large, the effect of biased estimation is clearly not significant. For small sample sizes, however, it does matter.

We now estimate the confidence interval of population parameters. The idea here is to sample the population, then compute a mean and standard deviation for it, and try to infer what this tells us about the mean and standard deviation of the population. If we define μ_s and σ_s as the mean and standard deviation of the sampling distribution of a statistic S, then we can expect the sampling distribution of S to be approximately Gaussian (assume that $N > 30$) and the actual sampled statistic S lying somewhere in the interval $\mu_s - \sigma_s$ to $\mu_s + \sigma_s$, or $\mu_s - 2\sigma_s$ to $\mu_s + 2\sigma_s$, or $\mu_s - 3\sigma_s$ to $\mu_s + 3\sigma_s$— about 68.3%, 95.5%, and 99.7% of the time, respectively.

In other words, we can be confident of finding μ_s in the interval $S - \sigma_s$ to $S + \sigma_s$ about 68.3% of the time; or in the interval $S - 2\sigma_s$ to $S + 2\sigma_s$ about 95.5% of the time; or in the interval between $S - 3\sigma_s$ to $S + 3\sigma_s$ about 99.7% of the time. We can say that we are 68.3% confident that the mean lies somewhere in the interval $S \pm \sigma_s$, that we have 95.5% confidence that the mean lies somewhere in the interval $S \pm 2\sigma_s$; and that we have 99.7% confidence that the mean μ_s lies somewhere in the interval $S \pm 3\sigma_s$. The relationship between confidence levels and σ levels is tabulated in Table 10-3.

Table 10-3 Confidence Levels Associated with σ Levels

Confidence Level (%)	σ Level z_c
50	0.6745
68.27	1.0000
80	1.28
90	1.645
95	1.96
95.45	2.00
96	2.05
98	2.33
99	2.58
99.73	3.00

The formula for computing the confidence interval associated with the mean is given by

$$\overline{X} \pm z_c \sigma_{\overline{x}}$$

If the sampling is from an infinite population or from a finite population but with replacement, the confidence limits in the estimate of the mean are specified by

$$\overline{X} \pm z_c \frac{\sigma}{\sqrt{N}} \qquad (10\text{-}49)$$

if sampling is without replacement from a population of finite size N_p:

$$\overline{X} \pm z_c \frac{\sigma}{\sqrt{N}} \left(\frac{N_p - N}{N_p - 1} \right)^{1/2} \tag{10-50}$$

The equations for confidence intervals for differences and sums of two statistics S_1 and S_2 are given by

$$(S_1 - S_2) \pm z_c \sigma_{S_1 - S_2} = (S_1 - S_2) \pm z_c \left(\sigma_{S_1}^2 + \sigma_{S_2}^2 \right)^{1/2} \tag{10-51}$$

$$(S_1 + S_2) \pm z_c \sigma_{S_1 + S_2} = (S_1 + S_2) \pm z_c \left(\sigma_{S_1}^2 + \sigma_{S_2}^2 \right)^{1/2} \tag{10-52}$$

provided that the distribution of S_1 and S_2 is approximately Gaussian.

By way of example, note that the confidence limits for the difference of two populations means in a case where the populations are infinite or are finite but with replacement are given by

$$\left(\overline{S}_1 - \overline{X}_2 \right) \pm z_c \sigma_{X_1 - X_2} = \left(\overline{X}_1 - \overline{X}_2 \right) \pm z_c \left(\frac{\sigma_1^2}{N_1} + \frac{\sigma_2^2}{N_2} \right)^{1/2} \tag{10-53}$$

The confidence interval for standard deviations of a normally distributed population is given by

$$S \pm z_c \sigma_S = S \pm \frac{\sigma}{\sqrt{2N}} \tag{10-54}$$

Occasionally we need to reference probable error; we define it here but retain the concept for reference. The 50% confidence limits of the population parameters corresponding to a statistic S are given by $S \pm 0.6745\sigma_s$. This quantity is known as the probable error of the estimate and may be worth memorizing.

10-10 SAMPLING IN THE SMALL

Earlier in the book we made use of the fact that there are simplifications in the formulas for computing a statistic when $N > 30$. Here we are concerned with cases when N is substantially less than 30, and the emphasis is on the determination of the statistics associated with small samples and on the distribution of those statistics. Specifically, we consider here the students t-distribution and the chi-square distribution.

The students t-distribution is defined by

$$t = \frac{\overline{X} - \mu}{s}\sqrt{N-1} = \frac{\overline{X} - \mu}{s/\sqrt{N}} \tag{10-55}$$

If we consider samples of size N selected from a Gaussian distribution with mean μ and if we compute t given the sample mean and sample standard deviation s, the sampling distribution for t can be obtained. This distribution is given by

$$Y = \frac{Y_0}{\left(1 + t^2/(N-1)\right)^{N/2}} = \frac{Y_0}{\left(1 + \dfrac{t^2}{\nu}\right)^{(\nu+1)/2}} \tag{10-56}$$

Here Y_0 is a constant depending on N and is such that the area under the t-distribution is 1. The constant $\nu = (N-1)$ is called the number of degrees of freedom. This distribution is called "students" t-distribution. Note that for large values of ν or $N(N>30)$ the curves closely approximate the Gaussian distribution:

$$Y = \frac{1}{\sqrt{2\pi}} e^{-t^2/2} \tag{10-57}$$

We can define the 95 and 99% confidence intervals by either computing the confidence intervals on the pocket calculator or using a table of t-distributions. Specifically, if $-t_{0.975n} + t_{0.975}$ are the values of t for which 2.5% of the area lies in each tail of the t-distribution, then a 95% confidence level for t is

$$-t_{0.975} < \frac{\overline{X} - \mu}{s}\sqrt{N-1} < t_{0.975} \tag{10-58}$$

Clearly, then, μ is expected to lie in the interval

$$\overline{X} - t_{0.975}\left(\frac{s}{\sqrt{N-1}}\right) < \mu < \overline{X} + t_{0.975}\left(\frac{s}{\sqrt{N-1}}\right) \tag{10-59}$$

with 95% confidence.

In general, we can represent the confidence limits for population means by

$$\overline{X} \pm t_c \frac{s}{\sqrt{N-1}} \qquad (10\text{-}60)$$

10-11 CHI-SQUARE

We now proceed to the chi-square random variable, which is defined by

$$\chi^2 = \frac{Ns^2}{\sigma^2} = \frac{\left(X_1 - \overline{X}\right)^2 + \left(X_2 - \overline{X}\right)^2 + \cdots + \left(X_N - \overline{X}\right)^2}{\sigma^2} \qquad (10\text{-}61)$$

The chi-square distribution is defined by

$$Y = Y_0(\chi^2)^{1/2(\nu-2)} e^{-(\chi^2/2)} = Y_0 \chi^{\nu-2} e^{-\chi^2/2} \qquad (10\text{-}62)$$

Here $\nu = N - 1$ is the number of degrees of freedom and Y_0 is a constant depending on ν such that the total area under the curve is 1. The chi-square distribution corresponding to various values of ν are shown in Figure 10-5. As was done with the normal and t-distributions, we can

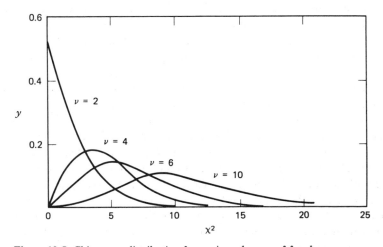

Figure 10-5 Chi-square distribution for various degrees of freedom.

define 95 and 99% or other confidence limits and intervals for chi-square by using a table of the chi-square distribution. In this manner we can estimate, within specified confidence limits, the population standard deviation σ in terms of the sample standard deviation s. If $\chi^2_{0.025}$ and $\chi^2_{0.095}$ have the values of χ^2 for which 2.5% of the area lies in each tail of the distribution, the 95% confidence interval is

$$\chi^2_{0.025} < \frac{Ns^2}{\sigma^2} < \chi^2_{0.975} \tag{10-63}$$

We can see that σ is estimated to be in the interval

$$\frac{s\sqrt{N}}{\sqrt{\chi^2_{0.975}}} < \sigma < \frac{s\sqrt{N}}{\sqrt{\chi^2_{0.025}}} \tag{10-64}$$

with 95% confidence.

10-12 CHI-SQUARE TESTING

The chi-square test is probably the most accepted test for determining whether there is significance between the observed and the expected in sampling a population. There are better tests. Once familiar with the chi-square test—the only one discussed in this book—the student and infrequent statistical analyst may wish to go on and examine the other tests as well.

It is only reasonable to expect that the results of a sampling of the population will not agree exactly with expected values. In fact, it is possible to partition a sample into a set of possible events E_1, E_2, E_3, $E_4,\ldots,E_n$, and then tabulate the observed frequency with which these events occur and the expected frequency from expectation analysis. For example, the distribution function for a particular experiment might be expected to be a Poisson distribution. After some hand analysis the expected value of the parameter λ in the distribution function is identified and the distribution itself is drawn. Then the interval of the random variable can be partitioned into subintervals and the probability that the random variable will occur in these intervals is computed on the basis of the expected distribution. These expected frequencies can then be tabulated for each event, as shown in Table 10-4. After completing the table by filling in the probabilities or frequencies of the observed events we are ready to perform the chi-square test. Specifically, we are ready to determine whether the observed frequencies differ significantly from the expected frequencies.

Table 10-4 One-Way Classification Table

Event	E_1	E_2	E_3	...	E_k
Observed Frequency	o_1	o_2	o_3	...	o_k
Expected Frequency	e_1	e_2	e_3	...	e_k

A definition of the difference between the observed and expected frequencies is given by the statistic χ^2:

$$\chi^2 = \frac{(o_1 - e_1)^2}{e_1} + \frac{(o_2 - e_2)^2}{e_2} + \dots + \frac{(o_k - e_k)^2}{e_k} = \sum_{j=1}^{k} \frac{(o_j - e_j)^2}{e_j} \tag{10-65}$$

Here the total frequency is N:

$$N = \Sigma o_j = \Sigma e_j \tag{10-66}$$

An alternate form of equation 10-65 commonly found in the literature is

$$\chi^2 = \Sigma \frac{o_j^2}{e_j} - N \tag{10-67}$$

Note that when the observed and expected frequencies agree, $\chi^2 = 0$. Obviously, the greater the difference between the two, the greater χ^2. The sample distribution of χ^2 can be closely approximated by the χ^2 distribution

$$Y = Y_0 \chi^{\nu-2} e^{-\chi^2/2} \tag{10-68}$$

when the expected frequencies get larger than 7. In fact, as the expected frequencies increase, the distribution more and more closely approaches the χ^2 distribution. Here the number of degrees of freedom is given by

$$\nu = k - 1 \tag{10-69}$$

when the expected frequencies can be computed without having to determine the χ^2 distribution parameters from the sample statistics themselves, and the number of degrees of freedom η is given by

$$\nu = k - 1 - m \tag{10-70}$$

if m population parameters are determined from the sample statistics.

The table of observed and expected frequencies can be extended to a table of k columns and h rows in which the observed frequencies occupy h rows and k columns. These are commonly called contingency tables. Corresponding to each observed frequency there is an expected or theoretical frequency. The χ^2 is then computed for the entire matrix, using the formula

$$\chi^2 = \sum_j \frac{(o_j - e_j)^2}{e_j} \qquad (10\text{-}71)$$

where the sum is taken over all the elements of the matrix made up by the contingency table. The sum of all observed frequencies is noted by N and is equal to the sum of all expected frequencies. As with the one-way classification table (Table 10-5) the distribution of the statistic for the $h \times k$ table is given very closely by the χ^2 distribution provided that the frequencies are not too small.

The number of degrees of freedom of this χ^2 distribution for $k > 1$ and $h > 1$ is given by

$$\nu = (h-1)(k-1) \qquad (10\text{-}72)$$

if the expected frequencies are not computed from the population sample. If they are, the degrees of freedom are given by

$$\nu = (h-1)(k-1) - m \qquad (10\text{-}73)$$

when m population parameters are determined from the sample statistics.

Usually expected frequencies are computed on the basis of a hypothesis we are trying to test. If the computed values of χ^2 are greater than some critical significance levels, we can conclude that the observed frequencies differ significantly from the expected frequencies. In this case we would reject the hypothesis being tested at the test level of significance. For example, if the value of χ^2 were to exceed the $\chi^2_{0.95}$, we would say the hypothesis is rejected at a 95% level, that is, there is a 5% chance that the rejection of the test could be wrong. Conversely, there then is a 95% chance that we are correct in rejecting the hypothesis.

In a similar manner, we would expect that, when χ^2 is approximately zero, the observed frequencies are in too close agreement with the expected frequencies. Obviously, the approach here is to also reject those cases that we consider to be in too close agreement, that is, when χ^2 is less than the expected value of χ^2 95% of the time, such as $\chi^2_{0.05}$.

The procedure we have just described assumes that we are going to compare results from frequencies determined with continuous distributions but applied to the discrete data associated with the finite samples. Yates has developed a correction factor to account for this difference. The corrected value of χ^2 is given by

$$\chi^2_{corrected} = \frac{(|o_1 - e_1| - 0.5)^2}{e_1} + \frac{(|o_2 - e_2| - 0.5)^2}{e_2} + \ldots + \frac{(|o_k - e_k| - 0.5)^2}{e_k}$$

$$(10\text{-}74)$$

This correction is made only when $\eta = 1$. For large samples, the corrected χ^2 approaches the uncorrected χ^2. The reason for the correction is that the uncorrected χ^2 can lead to significant errors near the critical values, that is, the low probability values of the χ^2 distribution. For small samples where the expected frequency is on the order of 7, it is better to use the corrected value of χ^2. In fact, some authors suggest the use of testing with both the corrected and uncorrected values of χ^2; if both lead to the same conclusion, the conclusion then is said to be unambiguously redefined.

There are a number of simple formulas for pocket calculator determination of χ^2 that use only the observed frequencies. For the χ^2 testing table shown in Table 10-5, χ^2 can be computed as follows:

$$\chi^2 = \frac{N(a_1 b_2 - a_2 b_1)^2}{(a_1 + b_1)(a_2 + b_2)(a_1 + a_2)(b_1 + b_2)} = \frac{N\Delta^2}{N_1 N_2 N_A N_B} \qquad (10\text{-}75)$$

where $\Delta = (a_1 b_2 - a_2 b_1)$. With Yates' correction

$$\chi^2_{corrected} = \frac{N(|\Delta| - N/2)^2}{N_1 N_2 N_A N_B} \qquad (10\text{-}76)$$

Table 10-5 A 2×2 Table of Observed Frequencies

	Event		
Event Group	I	II	Total
A	a_1	a_2	N_A
B	b_1	b_2	N_B
Total	N_1	N_2	N

Similarly, for a 2×3 table, χ^2 is given by

$$\chi^2 = \frac{N}{N_A}\left[\frac{a_1^2}{N_1} + \frac{a_2^2}{N_2} + \frac{a_3^2}{N_3}\right] + \frac{N}{N_B}\left[\frac{b_1^2}{N_1} + \frac{b_2^2}{N_2} + \frac{b_3^2}{N_3}\right] - N \quad (10\text{-}77)$$

Example 10-1   If 20% of the transistors produced by a process are defective, using the binomial distribution determine the probability that out of four transistors chosen at random, one, none, and at most two will be defective. Since the probability of a defective transistor is

$$P = .2$$

and of a nondefective transistor it is

$$q = 1 - p = .8$$

then

probability (1 defective transistor out of 4)

$$= {}_4C_1(.2)^1(.8)^3 = \frac{4!}{1!(4-1)!}(.2)^1(.8)^3 = .4096$$

probability (0 defective transistors) $= {}_4C_0(.2)^0(.8)^4 = .4096$

probability (2 defective transistors) $= {}_4C_2(.2)^2(.8)^2 = .1536$

Thus

probability (at most 2 defective transistors)

$$= \text{probability (0 defective transistors)}$$

$$+ \text{probability (1 defective transistor)}$$

$$+ \text{probability (2 defective transistors)}$$

$$= .4096 + .4096 + 1536 = .9728$$

Example 10-2   The mean weight of 500 engineers is 170 lb, and the standard deviation is 15 lb. Assuming the weights to be normally distributed, find how many engineers weigh between 139 and 175 lb and how many weigh more than 185 lb. Weights recorded as being between 139 and 174 lb can actually have any value from 138.5 to 174.5 lb, assuming that they are recorded to the nearest pound.

$$138.5 \text{ lb in standard units} = \frac{138.5 - 151}{15} = -2.10$$

$$174.5 \text{ lb in standard units} = \frac{174.5 - 151}{15} = 0.30$$

Then the number of engineers between 139 and 174 lb is (Figure 10-6)

(area between $z = -2.10$ and $z = 0.30$) = (area between $z = -2.10$ and z

$$= 0) + (\text{area between } a = 0 \text{ and } z = 0.30) = 0.4821 + 0.1179 = 0.6000$$

The number of engineers weighing between 139 and 174 lb is 500(0.6000) = 300.

Figure 10-6 Percent of engineers weighing between 139 and 174 lb.

Example 10-3 Find the probability of obtaining between 3 and 6 heads inclusive in 10 tosses of a coin by using the binomial distribution and the normal approximation of the binomial distribution.

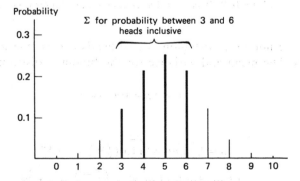

Figure 10-7 Discrete binomial distribution.

The binomial distribution gives (Figure 10-7)

$$\text{probability (3 heads)} = {}_{10}C_3\left(\tfrac{1}{2}\right)^3\left(\tfrac{1}{2}\right)^7 = \tfrac{15}{128}$$

$$\text{probability (4 heads)} = {}_{10}C_4\left(\tfrac{1}{2}\right)^4\left(\tfrac{1}{2}\right)^6 = \tfrac{105}{512}$$

$$\text{probability (5 heads)} = {}_{10}C_5\left(\tfrac{1}{2}\right)^5\left(\tfrac{1}{2}\right)^5 = \tfrac{63}{256}$$

$$\text{probability (6 heads)} = {}_{10}C_6\left(\tfrac{1}{2}\right)^6\left(\tfrac{1}{2}\right)^4 = \tfrac{105}{512}$$

Thus

probability (of getting between 3 and 6 heads inclusive)

$$= 15/128 + 105/512 + 63/256 + 105/512 = 99/128 = 0.7734$$

The probability distribution for the number of heads in 10 tosses of the coin is shown graphically in Figure 10-8. The required probability is the sum of the areas of the cross-hatched rectangles and can be approximated by the area under the corresponding normal curve shown dashed.

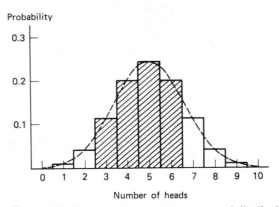

Figure 10-8 Data treated as continuous normal distribution.

Using the normal distribution, 3 to 6 heads can be considered as 2.5 to 6.5 heads. The mean and variance for the binomial distribution are given by

$$\mu = Np = 10\left(\tfrac{1}{2}\right) = 5$$

and

$$\sigma = \sqrt{Npq} = \sqrt{(10)\left(\tfrac{1}{2}\right)\left(\tfrac{1}{2}\right)} = 1.58$$

$$2.5 \text{ in standard units} = \frac{2.5 - 5}{1.58} = -1.58$$

and

$$6.5 \text{ in standard units} = \frac{6.5-5}{1.58} = 0.95$$

The probability of getting between 3 and 6 heads inclusive in the area between $z = -1.58$ and $z = 0.95$ under the normal distribution curve is

(area between $z = -1.58$ and $z = 0$) + (area between $z = 0$ and $z = 0.95$)

$$= 0.4429 + 0.3289 = 0.7718$$

which compares very well with the true value 0.7734 obtained using the binomial distribution.

Example 10-4   Measurements of the diameters of a random sample of 200 bearings made by a certain machine showed a mean diameter of 0.824 in. and a standard deviation of 0.042 in. Find (*a*) 95% and (*b*) 99% confidence limits for the mean diameter of all the bearings that will be made by this machine.

The 95% confidence limits are given by

$$\bar{X} \pm \frac{1.96\sigma}{\sqrt{N}}$$

which is approximately

$$\bar{X} \pm \frac{1.96\hat{s}}{\sqrt{N}}$$

In this example, $\bar{X} = 0.824$ in. and $\hat{s} = 0.042$ in. Thus

$$0.824 \pm 1.96 \left(\frac{0.042}{\sqrt{200}} \right)$$

$$0.824 \pm 0.0058 \text{ in.} = (0.8182, 0.8240)$$

with a 95% confidence interval.

The 99% confidence limits are given by

$$\bar{X} \pm \frac{2.58\hat{s}}{\sqrt{N}}$$

which is numerically equal to

$$0.824 \pm 2.58 \left(\frac{0.042}{\sqrt{200}} \right)$$

Thus

$$0.824 \pm 0.0077 \text{ in.} = (0.8163, 0.8317)$$

with a 99% confidence interval. Note that we have assumed the reported standard deviation to be the unbiased standard deviation $\hat{s}$. If the standard deviation had been s, we would have used $\hat{s} = s\sqrt{N/(N-1)}$ $= \sqrt{200/199}s$, which can be taken as s for all practical purposes.

Example 10-5 A small sample of 10 measurements of the length of a bolt give a mean of $\overline{X} = 4.38$ in. and a standard deviation of $s = 0.06$ in. Find the 95% and 99% confidence limits for the actual length.

The 95% confidence limits are given by

$$X \pm t_{0.975}\left(\frac{s}{\sqrt{N-1}}\right)$$

Since

$$\nu = 9$$

we can find

$$t_{0.975} = 2.26$$

from Table 10-6 or directly using the definition of the t distribution. Using $\overline{X} = 4.39$ and $s = 0.06$, we can be 95% confident that the actual mean will be included in the interval

$$4.38 \pm 2.26\left(\frac{0.06}{\sqrt{10-1}}\right) = 4.38 = 0.0452 \text{ in.}$$

The 99% confidence limits are given by

$$\overline{X} \pm t_{0.995}\left(\frac{s}{\sqrt{N-1}}\right)$$

Where $\nu = 9$ and $t_{0.995} = 3.25$. Then with 99% confidence we can expect the actual mean to be included in the interval

$$4.38 \pm 3.25\left(\frac{0.06}{\sqrt{10-1}}\right) = 4.38 \pm 0.0650 \text{ in.}$$

Now we work this problem, assuming that large sampling methods are

Table 10-6 Percentile Values (t_p) for the Single-Sided
t-Distribution with ν Degrees of Freedom

ν	$t_{0.995}$	$t_{0.99}$	$t_{0.975}$	0.95	$t_{0.90}$
1	63.66	31.82	12.71	6.31	3.08
2	9.92	6.96	4.30	2.92	1.89
3	5.84	4.54	3.18	2.35	1.64
4	4.60	3.75	2.78	2.13	1.53
5	4.03	3.36	2.57	2.02	1.48
6	3.71	3.14	2.45	1.94	1.44
7	3.50	3.00	2.36	1.90	1.42
8	3.36	2.90	2.31	1.86	1.40
9	3.25	2.82	2.26	1.83	1.38
10	3.17	2.76	2.23	1.81	1.37
11	3.11	2.72	2.20	1.80	1.36
12	3.06	2.68	2.18	1.78	1.36
13	3.01	2.65	2.16	1.77	1.35
14	2.98	2.62	2.14	1.76	1.34

valid, and compare the results of the two methods. Using large sampling methods, we obtain the 95% confidence limits

$$\overline{X} \pm \frac{1.96\sigma}{\sqrt{N}} = 4.38 \pm 1.96\left(\frac{0.06}{\sqrt{10}}\right) = 4.38 \pm 0.037 \text{ in.}$$

where we have used the sample standard deviation 0.06 as an estimate of σ. This is to be compared with 4.83 ± 0.0452 using small sample statistics.

Similarly, the 99% confidence limits are

$$\overline{X} \pm \frac{2.58\sigma}{\sqrt{N}} = 4.38 \pm 2.58\left(\frac{0.06}{\sqrt{10}}\right) = 4.38 \pm 0.049 \text{ in.}$$

as compared with 4.83 ± 0.0650.

In each case the confidence intervals obtained by using the small or exact sampling methods are greater than those obtained by using large sampling methods. The reason is that less precision is available with small samples than with large samples.

Example 10-6   Test scores of 16 students from one city showed a mean of 107 with a standard deviation of 10, while the same test scores of 14 students from another city showed a mean of 112 with a standard deviation of 8. Is there a statistically significant difference between the mean scores of the two groups at a 0.01 and a 0.05 level of significance?

If μ_1 and μ_2 denote population mean scores of students from the two cities, we will see if we can reject the hypothesis

$$H_0: \mu_1 = \mu_2$$

That is, there is no statistically significant difference between the groups.

The reader can easily show for himself that in the case of comparing two sample sets

$$t = \frac{\bar{X}_1 - \bar{X}_2}{\sigma\sqrt{1/N_1 + 1/N_2}}$$

where

$$\sigma = \sqrt{\frac{N_1 s_1^2 + N_2 s_2^2}{N_1 + N_2 - 2}} = \sqrt{\frac{16(10)^2 + 14(8)^2}{16 + 14 - 2}} = 9.44$$

Then

$$t = \frac{112 - 107}{9.44\sqrt{1/16 + 1/14}} = 1.45$$

On the basis of a *two-tailed test* at a 0.01 level of significance, we would reject H_0 if t were outside the range $-t_{0.995}$ to $t_{0.995}$. For $(N_1 + N_2 - 2)$ degrees of freedom $= (16 + 14 - 2) = 28$, this range is -2.76 to 2.76. It follows that we cannot reject H_0 at a 0.01 level of significance.

We would reject H_0 on the basis of a two-tailed test at a 0.05 level of significance if t were outside the range $-t_{0.975}$ to $t_{0.975}$. For 28 degrees of freedom this range is -2.05 to 2.05. Again we cannot reject H_0 at a 0.05 level of significance.

We conclude, therefore, that there is no significant difference between the scores of the two groups.

Example 10-7   The standard deviation of the heights of 16 male students chosen at random in a school of 1000 male students is 2.40 in. Find 95% and 99% confidence limits of the standard deviation for all male students at the school.

The 95% confidence limits are given by $s\sqrt{N}\ /\chi_{0.975}$ and $s\sqrt{N}\ /\chi_{0.025}$. For

$$\nu = 16 - 1 = 15 \text{ degrees of freedom}$$

$$\chi_{0.975}^{2} = 27.5$$

or

$$\chi_{0.975} = 5.24$$

and

$$\chi_{0.025}^{2} = 6.26$$

or

$$\chi_{0.025} = 2.50$$

Then the 95% confidence limits are $2.40\sqrt{16}\ /5.24$ and $2.40\sqrt{16}\ /2.50$, that is, 1.83 and 3.84 in. Thus we can be 95% confident that the population standard deviation lies between 1.83 and 3.84 in.

The 99% confidence limits are given by

$$s\sqrt{N}\ /\chi_{0.995} \quad \text{and} \quad s\sqrt{N}\ /\chi_{0.005}.$$

For

$$\nu = 16 - 1 = 15 \text{ degrees of freedom}$$

$$\chi_{0.995}^{2} = 32.8$$

or

$$\chi_{0.995} = 5.73$$

and

$$\chi_{0.005}^{2} = 4.60$$

or

$$\chi_{0.005} = 2.14$$

Then the 99% confidence limits are $2.40\sqrt{16}\ /5.73$ and $2.40\sqrt{16}\ /2.14$, that is, 1.68 and 4.49 in.

Thus we can be 99% confident that the population standard deviation lies between 1.68 and 4.49 in.

The number of degrees of freedom of a statistic are denoted by η, which is defined to be the number of independent observation samples minus the number of population parameters that must be estimated from the sample observations.

PART FOUR

THE PROGRAMMABLE
POCKET CALCULATOR

CHAPTER 11

THE PROGRAMMABLE
POCKET CALCULATOR

11-1 INTRODUCTION

As mentioned in the preface, the premise of this book is that the pocket calculator provides the analyst with a new dimension capability. The programmable pocket calculator, in the author's opinion, is yet another advance in pocket computing capability for the scientific analyst.

From the analyst's viewpoint, the most significant use of the pocket calculator may turn out to be as a teaching machine. The usual approach to learning a new discipline or a new technology involves four steps:

1. Studying the discipline in the textbook fashion.
2. Identifying or developing the mathematical tools that have been useful for solving the discipline's problems.
3. Working the "textbook" problems to learn the details and subtleties of the discipline through quantifying the problems with numbers, tables, graphs, and drawings.
4. At least two to three years of application of the mathematical models by working in the discipline itself.

The last step in this process takes one from the textbook-type knowledge to actual knowledge of the discipline itself. Here, the mathematical models are usually more complex, requiring many subtle considerations. For example, Fourier analysis as studied in the textbooks is often confined to Fourier series and Fourier integral representations of continuous functions defined on the entire domain of the reals. In practical harmonics analysis, the functions are usually finite in length and the data are usually sampled, requiring "window carpentry" filtering of the data prior to conducting the harmonic analysis. These "practical problems," while discussed in textbooks, are rarely given the consideration that they require in actual data handling problems.

With the advent of the programmable pocket calculator, the analyst now has libraries available for many disciplines other than his own. These libraries have been developed by persons highly experienced in their disciplines. By merely securing the standard library for a specific discipline, the analyst who has a programmable pocket calculator can, in a matter of weeks, become familiar with the programs and mathematical tools used in the discipline. Having acquired experience with practical programs, he can then focus his attention on the problems in his discipline and how to use the programs to solve them. In a sense the learning process is reversed. The analyst begins with the ability to numerically evaluate problems with which he is only vaguely familiar. These are practical problems, however, and involve mathematical models that have passed the test of time in practical analysis. Under the guidance of a person experienced in this new discipline, the learning process is fast.

The programmable pocket calculator is a good teaching machine also in that it is portable and the learning can be conducted in the comfort of one's own home. In the past, the numerical evaluation of most practical engineering problems was usually done on either a digital computer or a programmable desk-top calculator at work (provided that the analyst could justify his request for a budget to run the computers). Now the analyst can study even the more complex aspects of any given problem or discipline at home, where most of us do our homework anyway. Furthermore, the analyst learns more quickly now because he spends most of his time thinking and deriving, with a minimum of effort (stroking the key strokes) on numerical evaluation.

The next most important capability that the pocket calculator brings to the scientific analyst is the iterative computation of numbers and the preparation of extensive tables and graphs (involving many point pairs) for a more extensive set of problems than could be handled on the nonprogrammable pocket calculator. For the many consultants and small engineering organizations that do not have a computer facility, the programmable pocket calculator can bring to each member of the staff tremendous computing power.

Scientists and engineers usually do not compute, but develop the formulas that are used to compute the numbers and thus provide the insight to solve problems. The pocket calculator allows the analyst to begin with a top-level mathematical model of his process and refine it very quickly by testing the model numerically. Here he develops a system of equations that he thinks will describe a process or solve a problem and uses the pocket calculator to numerically evaluate the equations so that the results can be compared with what is observed about the system. The analyst judges the degree to which the model can satisfactorily predict the behavior of the process. Any major discrepancies lead to revisions and improvement in the model. Thus, while Newton's laws and Lagrange's equations generate the equations of motion of a process that, when solved, will predict its behavior, the numerical comparison between the actual observed behavior and the predicted behavior may indicate that certain elements are left out of the models. In the Newtonian formulation of mechanics this would lead to more comprehensive free body diagrams to better understand the system and thus to derive better mathematical models. In the Lagrangian formulation of mechanics it would lead to the development of a more refined Lagrangian which would have more energy terms to account for the additional elements in the system.

We see, then, that the pocket calculator does not improve the *"method"* for generating the equations of motion but helps to improve the mathematical model to which the methods are applied. This development of mathematical models using numerical testing is a convenient and fast operation with the programmable pocket calculator. There is no waiting for a batch-processed computer run to be made to get the data for improving the model. There are no "charge numbers" or budget required to permit the analyst to use the computer. On the programmable pocket calculator the cost of the run is in the "noise" of the electric bill. Finally, when an acceptable mathematical model is developed, the analyst can transfer the model to a magnetic tape strip and store the model for future use—a convenient means for conserving the energy spent preparing the mathematical model. Furthermore, the key data used in the analysis of a problem can also be stored on the magnetic tape for future reference.

Thus the programmable pocket calculator also provides an effective means of documenting an analysis. The analyst can collect a magnetic tape library at relatively small expense, requiring minimal storage space at relatively low cost.

Finally, the programmable pocket calculator provides the engineer with portable low-cost computing power for use in the field, in his car, at his customer's location, or in the convenience of his home.

11-2 HARDWARE CONSIDERATIONS

The programmable pocket calculator has the following parts:

1. The arithmetic unit, that is, the combination of registers that perform the arithmetic.

2. The memory, which stores numbers and instructions as programmed from the keyboard or from stored programs on magnetic tape.

3. The firmware associated with the calculator, that is, its "hard wired" programs and instruction set that are already built into the calculator.

Calculators perform numerical calculations only, as opposed to computers, which are alpha-numeric data processors. Today's programmable calculator, and any that might be expected in the near future, will only be limited in the fact that they perform numerical calculations and not alpha-numeric operations. Apart from this, the pocket calculators are similar to digital computers. Specifically data can be stored in memory, recalled to the arithmetic registers or arithmetic unit, and processed and restored in memory following the sequence of preprogrammed operations. An essential and interesting difference between the typical calculator and its digital computer counterpart is that many calculators operate in decimal rather than binary arithmetic. The reason is that decimal arithmetic involves less electronics for the special-purpose calculators than would conversion from decimal to binary and back again, as on general-purpose digital computing machines. Memories therefore are often set up in integer multiples of 10, as is the number of registers in the computing machine. For example, certain pocket calculators have one constant storage register and four arithmetic registers, in which the register arithmetic is performed. Ten additional storage registers are available in an advanced model of the basic calculator. The HP-65, has 100 programmable steps that can be put into memory.

The memory in most programmable calculators can be expected to be a set of registers in conjunction with the operating stack for performing register arithmetic and for scratch-pad storage during the execution of a program. From the standpoint of memory for storing numbers, there are only the registers in the stack plus the scratch-pad registers for number storage and manipulation. For example, if there are 9 scratch-pad registers and 4 stack registers, there are 13 total storage locations for storing numbers generated by a program. There is, however, memory for integer multiples of 10 keyboard instructions. For example, in the HP-65, 100 instructions can be stored in the calculator for sequential operations. That is, a program of 100 *key strokes* on the keyboard of the machine can be stored and executed automatically. Though numbers can be programmed into the calculator, using the 100 storable key strokes is relatively inefficient. Instead, the numbers can be input into the scratch-pad memory directly as opposed to inputting a 13-digit number into memory with 13 key strokes. This is perhaps the only important distinction between programmable pocket calculators and the standard digital computer. The programmable pocket calculator can be expected to store about 100 to

100,000 *key strokes*, not 13-*digit numbers*. Thus when we say that a digital computer has 32*k* 16-bit words or that a desk-top calculator has 4*k* 12-bit words, for the pocket calculator we say that it can store 100 or so *key strokes*. This might seem somewhat limiting, but actually most pocket calculator problems involve fewer than 100 key strokes. With a 100-stroke memory capability we can evaluate rather advanced mathematical functions and program fairly sophisticated iterative procedures for solving difficult problems.

The speed with which the pocket calculator processes the 100 instructions varies from calculator to calculator. From 10 to 1000 instructions per second can be expected from present-day pocket calculator electronic circuitry.

11-3 FIRMWARE

The firmware consists of the instruction set built into the pocket calculator and "*called*" from its keyboard. The basic instruction set usually contains all of the functions on the keyboard of the scientific calculator and a set of special functions associated with the programming aspect of the programmable pocket calculator. These include the following:

1. The GO-TO instruction. This instructs the calculator to perform the instruction at the *n*th step in the stored program. Thus GO-TO 50 would tell the computer to perform the instruction at the fiftieth step in the program.

2. The JUMP instruction. This instructs the calculator to jump the next two steps. It is expected that this instruction will be a natural part of all programmable calculators, and the two steps that are skipped are usually GO-TO type instructions. Thus the JUMP instruction with the GO-TO instruction permits the calculations to be looped iteratively in the computer program.

3. The DECREMENT AND JUMP ON ZERO instruction. This instruction, which can reasonably be expected in programmable calculators, examines the contents of one of the scratch-pad storage registers. If the register is not zero, it decrements the register by 1 and continues. When the register is zero, it will perform the JUMP operation.

4. The LOGICAL or TEST FLAG instruction. The flag can be set equal to 1 or zero, thus controlling the data flow in a calculator program, based on whether the flag is 1 or zero. Usually, the test flags or Booleans can be set manually on the keyboard or, since it is a keyboard instruction, with the program.

5. The STOP instruction. This is an instruction to stop the program.

6. The TEMPORARY STOP or RUN/STOP instruction. The calculator is told temporarily to stop, usually for the purpose of data input or data output.

Other keyboard instructions that can be expected to be found on the typical programmable pocket calculators are the DELETE function, the NO-OP function, and the SINGLE-STEP function. The SINGLE-STEP function permits the program to be processed or reviewed a single step at a time. This is for the purpose of debugging the program and examining or modifying the program by stepping up to the location in the instruction sequence that is to be modified or changed and the DELETE function instruction used to delete the previously programmed instruction, leaving it available for reprogramming. Finally, the NO-OP function can be used to fill memory with an instruction not to perform an operation. In this way, the remaining steps of a program can be safeguarded against accidental programming of the instruction sequence with undesirable program steps.

The firmware in a programmable pocket calculator can also include a keyboard for performing user-defined functions—functions that are programmed in a normal manner by a sequence of key strokes telling the calculator how to execute the function. Of the firmware just discussed, only the latter uses part of the programmable memory; the former functions are part of the keyboard sequence and thus are designed into the electronics of the calculator.

11-4 SOFTWARE

The software in programmable calculators is usually a magnetic tape strip, a magnetic card strip, or a tape cassette that is used to both read in and read out data and instructions from- or -to the memory of the calculator. It can be expected that manufacturers will provide preprogrammed software for performing analysis for many disciplines. In fact, it is precisely this software that permits a single pocket calculator to be programmed to perform special-purpose calculations in many disciplines. In a discussion with the Chief Engineer on the HP-65 Program, Chung Tung, the author was informed that it was precisely this motivation that led Hewlett-Packard to develop the HP-65, the first of the programmable pocket calculators.

The software associated with any pocket calculator would usually be developed so that problems involving more than 100 instruction sets and requiring more than the scratch-pad storage provided by the stacks plus scratch-pad memory can be programmed on a series of mag tapes or mag card strips. In attempting to see how far this process could be carried, the author programmed an 11-card sequence on the HP-65 for executing the lateral and vertical channels of an automatic landing system simulation which involves the numerical integration of a 14th-order continuous dynamic process, including saturation limits and other hard-stop non-linearities. While somewhat impractical to use, it does point out the flexibility of this method when general-purpose computing machines are not available—when the calculation is required in the field or when research is being conducted away from the computer center.

A similar procedure can be used for storing data in excess of the 100 key strokes plus the limited scratch-pad storage available in the pocket calculator.

11-5 PROGRAMMABLE POCKET CALCULATOR TECHNIQUES

The basic procedure for solving a problem on the programmable pocket calculator is as follows:

1. *Definition of the problem.* The generic types of problem that are conveniently solved on the pocket calculator are data processing (which includes interpolation, extrapolation, and filtering), the numerical evaluation of functions, the solution to systems of equations (whether algebraic or differential), the simulation of continuous processes, the frequency-domain analysis of data, and the statistical analysis of data. All these topics are covered in this book.

2. *Preparation of a math flow of the sequence of key strokes required.* For this the equations for solving the problem must be determined and the sequence of key strokes to numerically evaluate the equation must be worked out in a form that can be solved explicitly, implicitly, or by a combination of both. The preparation of the math flow will, by definition, identify the control operations for automatic execution of the key strokes.

3. *Programming of the calculator by keying in the key stroke sequence, including control operations.* Once the program is stored in memory, it is useful to load it onto a mag-tape strip so as not to inadvertently destroy the program. It is reasonable to expect that programmable calculators will have an ERASE BEFORE WRITE tape load function. Thus reprogramming or redefining the program or modifying the program can be restored on the same mag-tape strip or cassette by simply reloading the program on the mag-tape strip.

4. *Verification and checking of the program* by tests with all numerical values set equal to zeros, 1, or a single sequence of numbers that permit testing the program and its loops.

5. *Running the sequence automatically* with the actual problem data.

As an example, consider the problem of analyzing a low-pass filter. Figure 11-1 shows the three steps in the mathematical modeling process. First, the physical block diagram model of the process is drawn, including all of the hardware elements, the system inputs, and the system outputs. In this particular case we have a high-gain amplifier with impedance networks on the forward and feedback loops, which result in the passage of the low-frequency components of the input signal to the output while attenuating the high-frequency components. The mathematical block diagram of this filter in Laplace transform notation is the second step in the modeling process shown in the figure. The frequency response of this filter can easily be determined by replacing S with $j\omega$ and computing the transfer function of the filter algebraically. For the more difficult problem of determining the time-domain response of the filter to arbitrary forcing functions, it is necessary to prepare the differential equation that models this physical program. This is the third step in the modeling process shown at the bottom of Figure 11-1. The next step is to prepare a math flow for the low-pass filter mathematical model.

The math flow visualizes the way in which the problem is intended to be solved on the programmable pocket calculator. As shown in Figure 11-2,

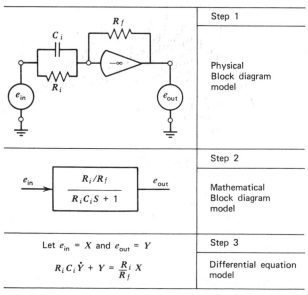

Figure 11-1   Low-pass filter mathematical model. $S = $ Laplace transformation operator.

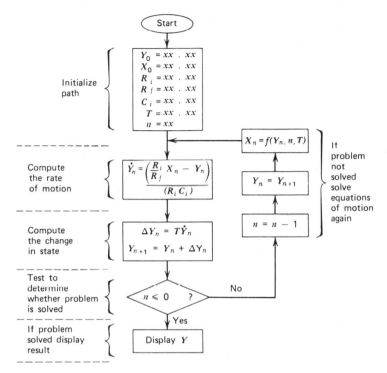

Figure 11-2 Math flow of low-pass filter mathematical model.

the first task is to initialize the problem with the state variables and parameters in the problem, together with the coefficients involved in the numerical integration process for solving the differential equations of motion of the low-pass filter and the control variable that is used to determine the computing path within the math flow. In this case the initial state vectors are the initial values of the filter input and output, the parameters are the resistance and capacitance of the filter, the parameter associated with the numerical integration process is the integration step size, and the control variable is the number of steps that we will take through the system of differential equations to compute the filter's response at a time nT.

After traversing the initialization path, the calculator is programmed to compute the rate of change of motion of the filter output. On computing the rate, the next step in the math flow is to compute the new value of the filter output, which is then followed by a test to determine if the calculations are to be stopped. In this example we ask if 100 passes through the system of equations have been taken. When the answer is yes, results are displayed in the display register. If the answer is no, another pass through the system of equations of motion is made (after renaming the variables and computing the forcing function for the next step, of course). In this example we simplify the situation by examining only the step response of the filter. The calculation of the forcing function in the closed-loop feedback process then is not necessary. This has no effect on the general nature of this discussion, since computing the forcing function is a straightforward process when it is necessary.

Table 11-1 illustrates the third step in the problem-solving procedure—preparing a key stroke sequence that can be programmed on the pocket calculator. In this example, the HP-65 calculator was used and some attention must be given to the details of its implementation on the HP-65

Table 11-1 Preparation of a Key Stroke Sequence

Math Flow	Math Flow Key Stroke	Function
Initialization Path	LBL-A	Label the initialization path A
	R/S	Stop then input and store y_0
	STO-1	
	R/S	Stop then input and store x_0
	STO-2	
	R/S	Stop then input and store R_i
	STO-3	
	R/S	Stop then input and store R_f
	STO-4	
	R/S	Stop then input and store C_i
	STO-5	
	R/S	Stop then input and store T
	STO-6	
	R/S	Stop then input and store n
	STO-8	
Identify loop closure points	LBL-1	Label this step "1"
Compute rate	RCL-2	Recall x
	RCL-3	Recall R_i
	RCL-4	Recall R_f
	÷	R_i/F_f
	×	$(R_i/R_f)x$

to understand this step of the procedure for solving problems on the programmable calculators. Though this key stroke sequence was programmed on the HP-65, it is typical of the key stroke sequences one would expect to encounter on most programmable pocket calculators.

The first 15 steps shown in Table 11-1 are the steps taken along the initialization path. The sequence begins by labeling the initialization path A to distinguish it from the normal feedback path which begins at step 16 and is labeled 1 (it identifies the point at which the feedback path loop closure occurs). The steps 2 through 15 involve automatically stopping to input a number on the keyboard and then manually starting the program again to store the keyboard number in memory. For example, step 2 stops the automatic program sequence so that the variable Y_0 can be input on

Table 11-1 (*Continued*)

| | Math Flow | |
Math Flow	Key Stroke	Function
Compute rate	RCL-1	Recall y
	CHS	$-y$
	$+$	$(R_i/R_f)x - y$
	RCL-3	Recall R_i
	RCL-5	Recall C_i
	$\times$	$R_i C_i$
	$\div$	$\dfrac{(R_i/R_f)x - y}{R_i C_i} = \dot{y}$
Compute state	RCL-6	Recall T
	$\times$	$\Delta y = T\dot{y}$
	RCL-1	
		$x = y + \Delta y$
	$+$	
	STO-1	$y_n = y_{n+1}$
Test for problem being solved	DSZ	Test register 8 for zero. If zero, skip next step and proceed
Go through equations of motion again if not solved	GTO-1	If R-8 not zero then go to step "1" and decrement R-8 by one and go to 1
Display results	R/S	Display y

the keyboard and then stored in location 1 (step 3) when the RUN-STOP key is again stroked (step 2). When operating, the computer will automatically progress to the first RUN-STOP (which is step 2), will stop, the variable Y_0 is manually input through the keyboard into the display register, and when the RUN-STOP key is stroked, the variable Y_0 will be stored in memory register 1 (step 3). The computer will then automatically proceed to step 4 where it will stop to await keyboard input of X_0 and the associated RUN-STOP key stroke to allow the program to proceed to step 5 which is to store the contents of the display register in memory register 2. This procedure of initializing the program continues until step 15 where the loop closure point is identified by labeling that particular step as step 1.

The next 12 steps compute the filter output rate according to the differential equation shown in Figure 11-2. Updating the filter output is done in the next five steps of the program. Then a test is made to see if 100 passes through the equations of motion have been completed. This is the step beginning with the key stroke DSZ, which means "decrement and skip on zero." The function of the DSZ is to test register 8 for zero. If register 8 is zero, the next step in the program will be skipped. In this example if register 8 is zero, the calculator will jump over the GO-TO instruction and go immediately to the RUN-STOP instruction where the program will stop and the latest value of Y will be displayed in the display register. If, however, the contents of register 8 are not zero, then the program will not skip the GO-TO instruction—it will go to the step labeled 1 and simultaneously the contents of register 8 will be decremented by 1. In Hewlett-Packard's implementation of the DSZ function, register 8 is used for the contents of the number of steps to be made in an iterative procedure. For other pocket calculators, it can be expected that other implementations of this DSZ function can be made. They all have one thing in common, however: the decrementing of *some* register and skipping the next step if the register's contents are zero. If it is not zero, the next step will not be skipped; however, the contents of the test register will be decremented by 1.

The careful reader will remember that the author recommends terminating an iterative procedure based on a test of the number of iterations if at all possible. Again, the reason is that for many iterative procedures an estimate of the number of steps to solve the problem can usually be made. This solution often gives insight into the convergence properties of the problem, which is helpful in establishing confidence in any result. For these problems where estimates of steps that should give the solution are known, the DSZ function is a "natural" test procedure and thus is particularly important in pocket calculator analysis.

Table 11-2 shows three *static check cases* used to check this program. Simple static tests of a program can often be made with numbers that are

Table 11-2 Preparation of a Check Case

Check Case 1	Check Case 2	Check Case 3
Let $Y_0 = 0$	$Y_0 = 1$	$Y_0 = 1$
$X_0 = 1$	$X_0 = 1$	$X_0 = 0$
$R_i = 1$	$R_i = 1$	$R_i = 1$
$R_f = 1$	$R_f = 1$	$R_f = 2$
$C_f = 1$	$C_f = 1$	$C_f = 3$
$T = 1$	$T = 1$	$T = 2$
$n = 1$	$n = 1$	$n = 1$
Then $\dot{Y} = 1$	$\dot{Y} = 0$	$\dot{Y} = -\frac{1}{3}$
$\Delta Y = 1$	$\Delta Y = 0$	$\Delta Y = -\frac{2}{3}$
$Y_n = 1$	$Y_n = 1$	$Y_n = \frac{1}{3}$
Display $Y_n = 1$	Display $Y_n = 1$	Display $Y_n = \frac{1}{3}$

quite unlike the physical characteristics of the process being studied. In this particular set of check cases, only zeros and 1 were used in the first two cases and zero through 3 for the third. It is important to develop dynamic check cases by either using an alternate means to solve the problem or a predetermined analysis of a simplified version of the problem. This is not shown as our straightforward example. The material covered in Chapter 7 is an example of the numerical methods that can be used to generate an independent check case on the dynamics of the solution of problems involving this type of differential equation. The idea would be to compare the results of a solution generated with a recursion formula with those of the solution generated here by using Euler numerical integration.

Table 11-3 illustrates the fifth step in the problem-solving procedure where the computer makes 100 passes through the equations to compute the filter's response at 1 second, with the filter design parameters being varied. The reader may be interested to know that the material for this example was developed and programmed and the sequence of solutions was run in approximately 17 minutes. The static check cases were run in 20 seconds, and the three 100-step solutions were run in approximately 3 minutes.

Further illustrations of the power of the programmable pocket calculator to solve problems are given in Chapter 12, where optimization problems with the penalty function method for handling equality constraints are programmed and example solutions are run to exemplify the calculator's use for this type of analysis.

Table 11-3 Running of the Automatic Sequence to Study the Unit Step Response of the Filter

Examples of 100-Step Solution Cases at $T = 0.01$			
Time	$R_i = 1\ M\Omega$ $R_f = 1\ M\Omega$ $C_i = 1\ \mu f$	$R_i = 2\ M\Omega$ $R_f = 2\ M\Omega$ $C_i = 1\ \mu f$	$R_i = 2\ M\Omega$ $R_f = 1\ M\Omega$ $C_i = \frac{1}{2}\ \mu f$
1 second	$y = 0.63396766$	$y = 0.39422956$	$y = 1.26793532$

It is worth pointing out that as the calculator is programmed, the key strokes are displayed according to their row-column location on the keyboard. For example, if the key that is at the intersection of the third column of keys and the second row of keys is depressed, a "32" is displayed in the register window. In this way, the programmer can monitor the programming of the process to ensure that the desired program is being stored in memory. This is also used in conjunction with the single-step key to review a program that is already in memory and, when necessary, to single-step up to the point where a change is to be made.

Relational tests that are not used in this example, but which can be expected in programmable pocket calculators, include those of whether a register is greater than, equal to, or less than the contents of another register. In the Hewlett-Packard 65 implementation, the relational tests are conducted in conjunction with the ninth memory register. As mentioned before, with the DSZ function it can be expected that other implementations will be available in other programmable pocket calculators.

Finally, a point well worth making is that in the preparation of any computer program on any programmable calculator (where there is more than enough memory for the problem), it is advisable to include additional RUN-STOP operations in long programs to display intermediate results while writing and checking the problem. When the program is finally checked out, the unwanted stops can be deleted. The deletion procedure is simply to single-step to the RUN-STOP and then use the DEL instruction to eliminate the RUN-STOP instructions used for checkout purposes.

11-6 METHODS OF ANALYSIS ON THE PROGRAMMABLE POCKET CALCULATOR

There are three basic types of numerical methods for solving problems on the programmable pocket calculator. In the explicit method the equations to be numerically evaluated are simply programmed on the calculator, thus eliminating the need for manually working out the sequence of key strokes to solve the problem. A manual optimization problem is a good example. Assume that the top-level cost model for some satellite programs takes the form*

$$C = n\left[30 + \left(\frac{M-1}{1.5}\right)30\right] + 4\left[30 + \left(\frac{M-1}{1.5}\right)30\right]$$

procurement   research, development,

refurbishment   launch costs

$$+ 0.4\left[30 + \left(\frac{M-1}{1.5}\right)30\right]\frac{T}{M} + \underset{\text{ground support}}{2T} + \frac{22T}{M}$$

tests, and engineering

which involves 69 key strokes for their numerical evaluation and the use of four scratch-pad storage locations. The program is tabulated in Table 11-4.

It is clear from the cost model that the mean mission duration of the satellite plays a dominant role in the cost equation. If the mean mission duration is small, the number of launches (T/M) is large, and the cost associated with each launch results in high total program cost. If the mean mission duration is large, the cost associated with the design and development of the satellite is large, which also leads to a high total program cost. Clearly, somewhere in between is a minimum total program cost. To determine it, we use a sequence of solution values for the cost equation, as shown in Table 11-4. It is apparent that a satellite mean mission duration of ~1.25 years minimizes the total program cost.

The preparation of Table 11-4 on the programmable pocket calculator involved 61 key strokes to program the calculator and 100 key strokes for data entry and manual program iteration. The entire procedure took 14 minutes, including checkout. When the table was manually prepared without using the programming feature of the calculator, the table took approximately 45 minutes to prepare. While the time saving shown here is typical of pocket calculator analysis, what is not shown (but what is equally important) is that had the total program cost model given unexpected or unexplained results that would have required its modification, only 3 minutes would have been required to incorporate the cost model

---

*n = number of satellites;
M = satellite mean mission duration;
T = total program lifetime; and
C = total program cost~millions of dollars.

Table 11-4 Total Program Cost for the XYZ Satellite Program

Number of Satellites	Program Duration (years)	Mean Mission Duration (years)	Total Program Cost (millions)
2	5	0.25	650
2	5	0.50	453
2	5	0.75	401
2	5	1.00	385
2	5	1.25	383
2	5	1.50	389
2	5	1.75	399
2	5	2.00	411
2	5	2.25	425
2	5	2.50	440
2	5	2.75	456
2	5	3.00	473
2	5	3.25	490
2	5	3.50	508
2	5	3.75	525
2	5	4.00	544
2	5	4.25	562
2	5	4.50	580
2	5	4.75	599
2	5	5.00	618

modifications and to prepare a new Table 11-4. With the programmable calculator, the modification would have been reprogrammed only for that part of the program where it was required.. The entire program need not necessarily be rewritten. Then only an additional 100 key strokes would have been needed to prepare another version of Table 11-4.

The second method of problem solving is the implicit method. An implicit equation is prepared and solved as discussed in Chapter 9 on determining zeros of a function. The procedure there would be to program the iterative procedure so that the solution to the implicit equation satisfies an error criteria established by the analyst.

The final procedure is neither implicit nor explicit. It is simply a brute-force search for the solution to an equation or system of equations by systematically testing regions where the solution is expected to exist and retaining only the value (or values) in the region that best satisfies the equation to be solved. Of the three methods, the latter is the most systematic, involving the least number of calculations and taking maximum advantage of the programmability feature of the pocket calculator. The only test that needs to be done is to determine whether the equation, when a solution is computed at a test point, is better satisfied with the test solution currently being used than with any previous test solutions. If it is, the new test point is stored in memory and the old one erased (or retained if it is desirable to monitor the convergence of the process). If it is not, the systematic search algorithm proceeds to the next test point, retaining the best previous test point. Of the iterative implicit and systematic search methods, the latter is the least efficient but involves the fewest programming steps, while the former method is more sophisticated, requiring logical tests and search algorithms, such as Newton's method.

From the analyst's viewpoint, the explicit mode of computer solution, where the analyst is involved in selecting the conditions to substitute into the computer program (manual iteration), is at best a gross procedure but requires only a few quick iterations, since the manual interaction will lead to a closing in on the gross solution fairly rapidly. The implicit method results in solutions that are difficult to develop with man-machine interaction because the precision with which the solution is to be determined is beyond the level at which the manual interaction can easily guess a better solution than a preprogrammed solution search algorithm (see Chapter 12).

Finally, the third method, while systematic and simple to program, results in the least efficient and least accurate solution to the problem. The accuracy can be refined through refined grids of possible solution values. The technique can be used for finding zeros of complex functions, such as those described in Chapter 9 on finding zeros of a function. It is a very practical and useful method when only a rough answer is required for a problem that takes a lot of key strokes to evaluate. Also, it is mentioned here as an example of the simplest form of problem solving available on the pocket calculator at a low programming overhead penalty.

CHAPTER 12

OPTIMIZATION

12-1 INTRODUCTION

No discussion of the programmable pocket calculator is complete without consideration of its optimization capabilities. The optimization problem has gained significance in engineering in the last few decades because it identifies the limit that practical design could approach if resources were unlimited. Practical engineering design usually is suboptimum design; the optimum is of vital importance nevertheless, since it identifies the ultimate design limit and optimum system capability.

Here we do *not* cover what is perhaps the key issue in any optimization work—the determination of what is to be optimized. Specifying precisely the payoff function in any systems analysis is a practical matter. It is perhaps the most difficult aspect of all systems analysis in that the computational analysis, once the payoff function has been identified, is almost a trivial matter in comparison to selecting the payoff function itself. In fact, commonly a number of payoff functions are identified and a system is optimized from a number of different viewpoints. The result is a group of optimized systems, which are studied to identify the most practical system.

Whatever the way in which optimization is applied in systems analysis, it is the analyst who must quantify the optimum system, from the standpoints of both its characteristics and its payoff. We therefore proceed to reexamine the fundamentals of optimization—for only the three simplest optimization problems: the parameter optimization without constraints, the parameter optimization with equality constraints, and the parameter optimization with inequality constraints. Though the simplest of the optimization problems, they are among the most frequently encountered. Also, they involve smaller programs than do the more sophisticated optimization problems and thus can often be handled on the pocket calculator. For these reasons, then, we reexamine the fundamental concepts of optimization to illustrate the optimization process by means of math flow and specific problems that can be programmed on the pocket calculator.

The approach used here is to first develop the mathematical concepts of these simple optimization problems and then discuss their numerical evaluation. The intention is to reacquaint the reader with the concepts of constraints, Lagrange multipliers, and the optimization terminology.

12-2 MAXIMA AND MINIMA

In most systems analysis the optimization problem amounts to maximizing the payoff function. This function is usually of the form "benefit divided by cost." As a system is developed and increasingly more money is spent on it, the benefits usually follow the law of diminishing returns. This is seen in Figure 12-1. It is also true that the benefits are usually accrued on a discrete basis as fixed amounts of money are spent on the system, rather than being continually accrued. It is apparent from Figure 12-1 that the benefit-to-cost ratio takes the shape shown in Figure 12-2. It is to the analyst's advantage, therefore, to maximize the cost-benefit curve or the payoff function in terms of the benefit-cost ratio.

From a more mathematical viewpoint, optimization involves either maximizing or minimizing a function $f(x_i)$. Specifically the objective is to identify those values of x_i that cause $f(x_i)$ to be a minimum or a maximum. Strictly speaking, we need only consider either the minimization or the

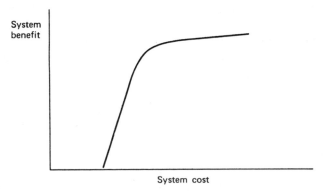

Figure 12-1 System benefit as a function of system cost.

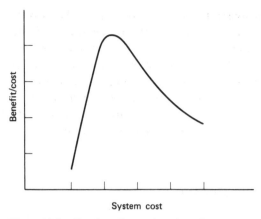

Figure 12-2   Cost benefit as a function of system cost.

maximization problem, but not both. The reason for this is that the values of x_i which maximize $f(x_i)$ also minimize $-f(x_i)$. The maximum of $f(x)$ occurs at the same place as does the minimum of $-f(x)$. We therefore discuss the optimization problem from the viewpoint of either extremum, but never both.

Perhaps the most familiar case in optimization is when the extremum of a single dependent variable is a function of a single independent variable. For a function of one variable, this means finding the point at which the derivative is zero and evaluating the function at that point. The value of the independent variable where the derivative is zero is only a necessary condition that the function be at an extremum; it is not sufficient. For example, a function can have a derivative equal to zero at a stationary point and not at a maximum or minimum. Thus it is necessary to check the second derivative to determine whether it too is at zero (a point of inflection). If it is not, the second derivative can be used to determine whether the extremum is a maximum or minimum depending on whether the second derivative is negative or positive. Thus the condition for the extremum of a single variable is

$$\frac{d}{dt}f(t)=0 \tag{12-1}$$

If $\dfrac{d^2}{dt^2}f(t)>0$, it is a minimum. (12-2)

If $\dfrac{d^2}{dt^2}f(t)<0$, it is a maximum. (12-3)

And if $(d^2/dt^2)f(t)=0$, it is a point of inflection.

For a two-variable function what is required is

$$\frac{d}{dx}f(x,y)=0 \qquad\qquad (12\text{-}4)$$

$$\frac{d}{dy}f(x,y)=0 \qquad\qquad (12\text{-}5)$$

Single variable optimization on the programmable pocket calculator was discussed in Chapter 11. It is worth mentioning again, however, that $f(x)$ can be conveniently programmed on the calculator and a search (manual or automatic) for the x that minimizes $f(x)$ can be quickly done. For complicated $f(x)$, the numerical value of the minimum $f(x)$ may be found more quickly on the calculator in this manner than by the analytical steps just outlined.

Optimization of functions of more than one variable force us to change notation at this point. In what follows, we use the notation of Bryson and Ho*:

$$x=\begin{bmatrix} x_1 \\ x_2 \\ x_m \end{bmatrix} = \text{parameter vector} \qquad\qquad (12\text{-}6)$$

We concern ourselves with the parameter optimization problems that involve finding the values of the m parameter $x_1,x_2,\ldots,x_m$ minimizing a payoff function that is a function of these parameters. We write the payoff function in the Lagrangian notation

$$L(x_1,x_2,\ldots,x_m)=L(x) \qquad\qquad (12\text{-}7)$$

The use of the programmable pocket calculator in solving optimization problems is presently limited to two or three dimensions, but it is quite useful for higher-dimension problems in computing "parts" of the problem as subroutines. In any case the pocket calculator permits optimization analysis of some complexity.

*See reference.

12-3 PARAMETER OPTIMIZATION WITHOUT CONSTRAINTS

If there are no constraints on x and if the function $L(x)$ has first and second partial derivatives, the necessary conditions for a minimum are

$$\frac{\partial L}{\partial x} = 0 = \frac{\partial L}{\partial x_i} \qquad (12\text{-}8)$$

and

$$\frac{\partial^2 L}{\partial x^2} \geqslant 0 \qquad (12\text{-}9)$$

Here we mean that the matrix whose components are $\partial^2 L / \partial x_i \partial x_j$ must have eigenvalues that are zero or positive. All x that satisfy $\partial L / \partial x = 0$ are called stationary points. When

$$\frac{\partial^2 L}{\partial x^2} > 0 \qquad (12\text{-}10)$$

at the stationary points x_s, $L(x_s)$ is at a local minimum. If $\partial^2 L / \partial x^2 = 0$ at $x = x_s$, it is not possible to establish whether the point is a minimum. Such a point is called a singular point.

12-4 PARAMETER OPTIMIZATION WITH EQUALITY CONSTRAINTS

A more general optimization problem is to find the values of the "control parameters" $u_1, \ldots, u_m$ that minimize a payoff function

$$L(x_1, \ldots, x_n; u_1, \ldots, u_m) \qquad (12\text{-}13)$$

where the n parameters $x_1, \ldots, x_n$ are determined by

$$f_1(x_1, \ldots, x_n; u_1, \ldots, u_m) = 0 \qquad (12\text{-}14)$$
$$\vdots$$
$$f_n(x_1, \ldots, x_n; u_1, \ldots, u_m) = 0 \qquad (12\text{-}15)$$

Now, let

$$x = \begin{bmatrix} x_1 \\ \vdots \\ x_n \end{bmatrix} = \text{parameter vector} \qquad (12\text{-}16)$$

$$u = \begin{bmatrix} u_1 \\ \vdots \\ u_m \end{bmatrix} = \text{control vector} \qquad (12\text{-}17)$$

$$f = \begin{bmatrix} f_1 \\ \vdots \\ f_n \end{bmatrix} = \text{constraint vector} \qquad (12\text{-}18)$$

Then the optimization problem is to find the vector u that minimizes

$$L(x, u) \qquad (12\text{-}19)$$

where the vector x is related to u according to the constraint equation

$$f(x, u) = 0 \qquad (12\text{-}20)$$

For a given optimization problem, the choice of which parameters to use as control parameters is not unique. The choice must be such that u determines x through the constraint equation.

A *stationary* point is one where $dL = 0$ for arbitrary du, while holding $df = 0$ (letting dx change as it will). Then we have

$$dL = L_x \, dx + L_u \, du \qquad (12\text{-}21)$$

and

$$df = f_x \, dx + F, du = 0 \qquad (12\text{-}22)$$

Equation 12-12 may be solved for dx:

$$dx = -f_x^{-1} f_u \, du \qquad (12\text{-}23)$$

By substitution, then, we have

$$dL = (L_u - L_x f_x^{-1} f_u)\, du \tag{12-24}$$

At the stationary point we see that for $dL = 0$ for any du

$$L_u - L_x f_x^{-1} f_u = 0 \tag{12-25}$$

These equations together with the constraint equations determine the u and x at stationary points.

Another technique is to adjoin the constraints to the payoff function by a set of n "undetermined Lagrangian multipliers," $\lambda_1, \ldots, \lambda_n$, as

$$H(x, u, \lambda) = L(x, u) + \sum_{i=1}^{n} \lambda_i f_i(x, u) = L(x, u) + \lambda^T f(x, u) \tag{12-26}$$

If we choose u (and thereby x through the constraint equations) so that $L = H$, and if we choose λ according to

$$\lambda^T = -\frac{\partial L}{\partial x} \left(\frac{\partial f}{\partial x} \right)^{-1}$$

then

$$\frac{\partial H}{\partial x} = 0$$

and

$$dL = dH = \frac{\partial H}{\partial u}\, du \tag{12-27}$$

Thus $\partial H / \partial u$ is the gradient of L with respect to u *while holding* $f(x, u) = 0$. At a stationary point, dL vanishes for arbitrary du; which can happen only if

$$\frac{\partial H}{\partial u} \equiv \frac{\partial L}{\partial u} + \lambda^T \frac{\partial f}{\partial u} = 0 \tag{12-28}$$

Hence a stationary value of $L(x, u)$ must satisfy the equations

$$f(x, u) = 0 \tag{12-29}$$

$$\frac{\partial H}{\partial x} = 0 \tag{12-30}$$

$$\frac{\partial H}{\partial u} = 0 \tag{12-31}$$

where

$$H = L(x, u) + \lambda^T f(x, u) \tag{12-32}$$

12-5 THE GRADIENT METHOD

When $L(x,u)$ and $f(x,u)$ are complex, numerical methods must be used to determine the values of u that minimize H. Perhaps the most commonly used numerical method is that of *steepest descent* for finding minima.

Gradient methods are iterative algorithms for estimating u, so as to satisfy the stationary conditions $\partial H/\partial u = 0$ (see Figure 12-3).

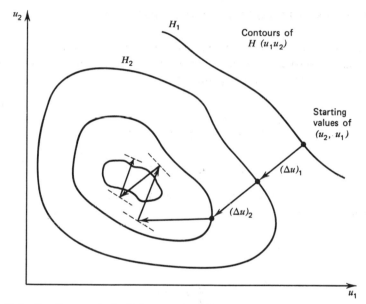

Figure 12-3 Gradient method. $(\Delta u)_1 = -k(\partial H_1/\partial u)$, $(\Delta u)_2 = -k(\partial H_2/\partial u)$. Note: search can overshoot when $-k(\partial H/\partial u) \gg 0$.

One procedure for using the gradient method is:

1. Estimate the values for u.
2. Compute x from $f(x,u)=0$.
3. Compute λ from $\lambda^T = -(\partial L/\partial x)(\partial f/\partial x)^{-1}$.
4. Compute $\partial H/\partial u = (\partial L/\partial u) + \lambda^T(\partial f/\partial u)$.
5. Revise the estimates of u by amounts $\Delta u = -K(\partial H/\partial u)^T$ (K is a positive scalar constant).
6. Iterate, using the revised estimates of u, until $(\partial H/\partial u)(\partial H/\partial u)^T$ is very small.

The gradient method for finding a minimum is a hill-descending technique. Starting with an initial guess of u, a sequence of changes Δu is made. At each step Δu is in the direction of the gradient $\partial H / \partial u$ whose magnitude gives the steepest slope at that point on the hill. The choice of K involves judgment to ensure that the linearized prediction will be accurate and the process will be efficient (i.e., will not require many iterations). K will usually be varied in the sequence of iteration when it is thought that the minimum is near.

12-6 COURANT'S PENALTY FUNCTION METHOD

Another numerical method for optimizing with either equality constraints or inequality constraints is the Courant penalty function method. Suppose that we wish to minimize $L(u)$ subject to

$$f(u) = 0 \tag{12-33}$$

For the penalty function method, we minimize

$$\bar{L} = L(u) + K \| f(u) \|^2 \tag{12-34}$$

subject to no constraints! Here K is large. If $\bar{L}$ attains a minimum at y_0, it is reasonable to expect that

$$f(u_0) \approx 0 \tag{12-35}$$

and

$$\frac{\partial L}{\partial u}(u_0) \approx 0 \tag{12-36}$$

Computationally, the penalty function method is easy to use and understand and has been used with great success in certain parameter optimization problems. The Courant penalty function method does not always work, however, because large values of K tend to make a long, narrow and deep depression in the field of $\bar{L}(u)$ with the stationary point at the bottom of the depression. The problem with this is that the gradient is more likely to be evaluated on the sides of the depression than the end of the depression. This will result in estimates of the stationary point that jump back and forth across the narrow depression instead of running down the length of the depression. For example, minimizing

$$L = (y_1 - 2)^2 + y_2^2 \tag{12-37}$$

subject to $y_1=0$ has $y_1=y_2=0$. The penalty function method minimizes

$$\bar{L}=(y_1-2)^2+y_2^2+Ky_1^2=y_2^2+\left[\left(y_1-\frac{2}{1+K}\right)^2\Big/\frac{1}{1+K}\right]+\frac{4K}{1+K} \quad (12\text{-}38)$$

Contours of constant $\bar{L}$ are ellipses with centers at $y_1=(2/1+K)$, $y_2=0$. Figure 12-4 shows contours of constant $\bar{L}$.

Inequality constraints can be conveniently handled by the penalty function method as well. The approach is straightforward and illustrated by the following example: Minimize $L(y)$ subject to the constraint $f(y)$

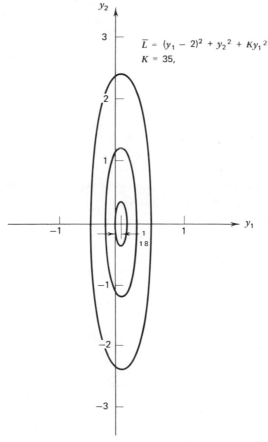

Figure 12-4 $\bar{L}$ contours created by large penalty function coefficients.

$\leqslant 0$. This problem is solved using the penalty function method by minimizing

$$\bar{L} = L(y) + K\mathbf{P}f(y^2) \tag{12-39}$$

where **P** is defined as

$$\mathbf{P} = \begin{cases} 1, & (f > 0) \\ 0, & (f < 0) \end{cases} \tag{12-40}$$

Examples of pocket calculator optimization with the penalty function and gradient methods for programmable pocket calculators follow.

Example 12-1 Find the stationary value of

$$L = \frac{1}{2}\left(\frac{x^2}{a^2} + \frac{u^2}{b^2}\right)$$

subject to the linear constraint

$$f(x,u) = x + mu - c = 0$$

(x is a scalar parameter and a, b, m, and c are constants).

The curves of constant L are ellipses, with L increasing with the size of the ellipse. The line $x + mu - c = 0$ is a fixed straight line. The minimum value of L satisfying the constraint is obtained when the ellipse is tangent to the straight line (Figure 12-5). Now

$$H = \frac{1}{2}\left(\frac{x^2}{a^2} + \frac{u^2}{b^2}\right) + \lambda(x + mu - c)$$

Thus the necessary conditions for a stationary value are

$$x + mu - c = 0, \qquad \frac{\partial H}{\partial x} = \frac{x}{a^2} + \lambda = 0, \qquad \frac{\partial H}{\partial u} = \frac{u}{b^2} + \lambda m = 0$$

These three equations for the three unknowns, x, u, λ, have the solutions

$$x = \frac{a^2 c}{a^2 + m^2 b^2}$$

$$u = \frac{b^2 mc}{a^2 + m^2 b^2}$$

$$\lambda = -\frac{c}{a^2 + m^2 b^2}$$

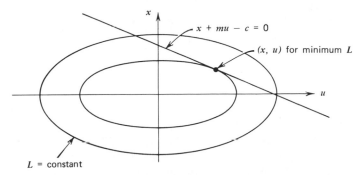

Figure 12-5 Example of minimization subject to constraint.

and the minimum value of L is given by

$$L_{min} = \frac{c^2}{2(a^2 + m^2 b^2)}$$

Note that

$$-\lambda = \frac{\partial J}{\partial c} = \frac{\partial J}{\partial f}$$

Example 12-2 The gradient optimization method (steepest descent)   Consider the problem of minimizing the function

$$L = (y_1 - 2)^2 + y_2^2$$

This problem, as mentioned in the last section of this chapter, is trivial from a practical viewpoint but instructive from a pocket calculator optimization viewpoint. By inspection we see that L is a minimum at $y_1 = 2$ and $y_2 = 0$. The gradient method is to seek the condition

$$\frac{\partial L}{\partial y} = 0 = \nabla L$$

We iteratively solve the implicit equation

$$y_{n+1} = y_n - k \nabla L(y_n)$$

until $\nabla L \approx 0$. Then $y_{n+1} \cong y_n =$ value of y that minimizes L. For this problem

$$\Delta y_1 = 2k_2(y_1 - 2)$$

$$\Delta y_2 = 2k_2 y_2$$

Then

$$y_{1_{n+1}} = y_{1_n} + \Delta y_{1_n}$$

$$y_{2_{n+1}} = y_{2_n} + \Delta y_{2_n}$$

The key stroke sequence for implementing the gradient optimization method on the HP-65 programmable pocket calculator (which is typical of the program for any pocket calculator) is shown in Table 12-1.

This sequence of key strokes was programmed in less than 1 minute. Now let us examine the types of numerical analyses that can be made with the pocket calculator. First consider the effect of k_2 on the first 10 steps of the process of finding the y_1 and y_2 that minimize L. In Table 12-2 $y_1 = 1 = y_2$ at the start ($L_0 = 2$). It is apparent that for the case of $y_1 = 1 = y_2$, $k_2 = 0.5$ leads to the best 10-step estimate of the optimum solution. The data for each step can be easily developed once the calculator is pro-

Table 12-1 Typical keystroke sequence for gradient optimization

Key Strokes		Comment
LBL-A		
R/S		Input y_1
STO-1		Store y_1 in register 1 (R-1)
R/S		Input y_2
STO-2	Input data	Store y_2 in R-2
R/S		Input K_2
STO-3		$K_2 \rightarrow$ R-3
R/S		Input N—the number of iterations
STO-8		$N \rightarrow$ R-8
LBL-1		Label this step "1"
RCL-1		Recall y_1
2		
–		$(y_1 - 2)$
STO-5		$(y_1 - 2) \rightarrow$ R-5
f^{-1}		
$\sqrt{\ }$	Compute L	$(y_1 - 2)^2$
RCL-2		y_2
f^{-1}		
$\sqrt{\ }$		y_2^2
+		$y_2^2 + (y_1 - 2)^2 = L$
STO-6		$L \rightarrow$ R-6

grammed. Thus for $y_1 = 1 = y_2$ and $k_2 = 0.05$ the sequence of solutions is as shown in Table 12-3. It appears that when $k_2 = 0.5$ the first step accidently puts the estimates of y_1 and y_2 right on the stationary points. This is seen by noting that y_1 and y_2 are exactly -1 for $k_2 = \frac{1}{2}$ and $y_1 = 1 = y_2$.

Now let us examine the effect of k_2 on the 10-step estimate of the stationary points when $y_1 = 3 = y_2$ (Table 12-4). Again $k_2 = 0.50$ results in the best 10-step estimate of the stationary points. The reason for this phenomenon might be thought to be that the initial conditions are an integer multiple of the step size 0.5 and the gradient is permitting the solution to fall precisely on the stationary points by accident. However, a

Table 12-1 (*Continued*)

Key Strokes		Comment
RCL-5		$(y_1 - 2)$
RCL-3		
×	Compute	$k_2(y_1 - 2)$
2	$\Delta y_1 = -\dfrac{\partial L}{\partial y_1} k_2$	
×		$2k_2(y_1 - 2)$
CHS		$-2k_2(y_1 - 2) = \Delta y_1$
RCL-1		y_1
+	Compute new y_1	$y_1 + \Delta y_1$
STO-1		$y_1 + \Delta y_1 \rightarrow$ R-1
RCL-2		y_2
RCL-3		k_2
×	Compute	$k_2 y_2$
2	$\Delta y_2 = -\dfrac{\partial L}{\partial y_2} k_2$	
×		$2k_2 y_2$
CHS		$-2k_2 y_2 = \Delta y_2$
RCL-2		y_2
+	Compute new y_2	$y_2 + \Delta y_2$
STO-2		$y_2 + \Delta y_2 \rightarrow$ R-2
DSZ		Number of steps $= N$?
GTO-1		If no, go to 1
RCL-6		If yes, recall L
R/S	Display	Display L
RCL-1	L, y_1, y_2	Recall y_1
R/S		Display y_1
RCL-2		Recall y_2
R/S		Display y_2
RTN		Return to the top of the program

Table 12-2 Tenth-Step Results for Various k_2

k_2	L_{10}	$y_{1_{10}}$	$y_{2_{10}}$
0.01	1.39027066	1.18292719	0.81707281
0.10	0.03602880	1.89262582	0.10737418
0.25	0.00000763	1.99902344	0.00097656
0.50	0.00000000	2.00000000	0.00000000
0.75	0.00000763	1.99902344	0.00097656
1.0	2.00000000	1.00000000	1.00000000
2.0	7.748409780×10^8	-5.9047×10^4	5.9049×10^4

Table 12-3 Iterative Gradient Optimization with $k_2 = 0.5$

Number of Iterations	L	y_1	y_2
0	2	1	1
1	0	2	0
2	0	2	0
.	.	.	.
.	.	.	.
.	.	.	.

Table 12-4 Tenth-Step Results for Various k_2 and $y_1 = y_2 = 3$

k_2	$(L)_{10}$	$(y)_{10}$	$(y_2)_{10}$
0.10	0.18014399	2.10737418	0.32212255
0.20	0.00101560	2.00604662	0.01813985
0.40	0.00000000	2.00000010	0.00000031
0.50	0.00000000	2.00000000	0.00000000
0.75	0.00003815	2.00097656	0.00292969
1.00	10.00000000	3.00000000	3.00000000

test of this hypothesis is but a few key strokes away, in that we can try the initial conditions $y_1 = \pi$ and $y_2 = 2\pi$ with keyboard entry. Then for $k_2 = 0.5$ we find

$$(L)_{10} = 0.00000000$$

$$(y_1)_{10} = 2.00000000$$

$$(y_2)_{10} = 0.00000000$$

where the sequence of estimates is as shown in Table 12-5. At this point it should be clear that $k_2 = 0.5$ is a unique value that causes the system of optimization equations to exhibit the peculiar property that the singular points are exactly determined on the first step of the iterative solution. This is precisely the case and it serves to illustrate the following key points:

1. Almost every system of equations has **unique numerical properties**. The analyst must keep ever alert for their discovery. Often it is possible to capitalize on them. The pocket calculator is an ideal means for this kind of research and exploration.

2. An understanding of the unique properties leads the analyst to a better understanding of the equations he is using. (It is left to the reader to determine why this simple system of equations has the property that the stationary points are *exactly* determined numerically when $k_2 = \frac{1}{2}$, no matter what the values of y_1 and y_2.)

In what follows we do not use the unique value of K_2 that exactly determines the singular points. To further illustrate the gradient method, let us use $k = 0.2$ and now consider the questions: How do we know we have reached the minimum L in a 10-iteration optimization? How can the analyst quickly gain confidence that the stationary point is not a local minimum of which there is a "deeper" minimum nearby? Mathematically

Table 12-5 Iterative Gradient Optimization with $k_2 = 0.5$ and Irrational Initial Conditions

Number of Iterations	L	y_1	y_2
0	40.74165139	π	2π
1	0	2	0
2	0	2	0
3	0	2	0
4	0	2	0
5	0	2	0

there is no guarantee that the gradient method will find THE global minimum of his function. Some practical things can be done, however, that give confidence in the end result of an optimization analysis, tieing all of these questions together and giving some plausible answers. Among them are the following:

1. Stopping the process for a fixed number of steps. This is the simplest criterion for terminating the iterative stationary point search process.

2. Selecting a new set of initial conditions and searching for the stationary point using the same number of steps.

3. Continuing (1) and (2) until convergence from all quadrants around the stationary point (initial one found) has been established.

4. Using the stationary point as the initial conditions and demonstrating stability of solution at the stationary point.

This procedure is quick, and will usually uncover the areas of concern if the results are different from initial conditions. Also, because more "samples" are available from the stationary point selection process, we tend to have more statistical confidence that we have indeed found the stationary point.

Returning to our optimization problem, we find the stationary point to be somewhat different when approached from different directions. Table 12-6 illustrates this point. It is apparent that L is a minimum in the near neighborhood of

$$y_1 = 2$$

$$y_2 = 0$$

Table 12-6 Ten-Step Gradient Optimization Results when Stationary Point Approached from Different Directions

y_{1_0}	y_{2_0}	L_{10}	$(y_1)_{10}$	$(y_2)_{10}$
5	4	0.002539	2.01813985	0.02418647
5	5	0.00345304	2.01813985	0.03023309
4	5	0.00294524	2.01209324	0.03023309
−4	4	0.00528112	1.96372029	0.02418647
−5	5	0.00751544	1.95767368	0.03023309
−5	4	0.00660140	1.95767368	0.02418647
−5	−4	0.00660140	1.95767368	−0.02418647
−5	−5	0.00751544	1.95767368	−0.03023309
−4	−5	0.00619516	1.96372029	−0.03023309
4	−4	0.0060203120	2.01209324	−0.02418647
5	−5	0.00345304	2.01813985	−0.03023309
4	−5	0.00294524	2.01209324	−0.03023309

The 100-step iteration bears this out, resulting in

$$L \approx 0$$

$$y_1 = 1.98790677$$

$$y_2 = 0.00000000$$

Example 12-3. The penalty function method Let us consider the problem of minimizing the function

$$L = (y_1 - 2)^2 + y_2^2$$

subject to the constraint

$$y_1 = 0$$

This problem, as mentioned before, is trivial from a practical viewpoint but is instructive in regard to pocket calculator optimization. Using the Courant penalty function method, we form the auxiliary function

$$\bar{L} = (y_1 - 2)^2 + y_2^2 + k_1 y_1^2$$

Our objective is to minimize this new function, using the programmable pocket calculator. For this example we employ the gradient method of optimization to merge the learning of both methods. Here

$$\frac{\partial \bar{L}}{\partial y} = \begin{cases} \dfrac{\partial \bar{L}}{\partial y_1} = +2[(1 + 2k_1)y_1 - 2] \\ \dfrac{\partial \bar{L}}{\partial y_2} = 2y_2 \end{cases}$$

Then

$$\Delta y = \begin{cases} -k_2 \dfrac{\partial \bar{L}}{\partial y_1} = -2[(1 + 2k_1)y_1 - 2]k_2 = \Delta y_1 \\ -k_2 \dfrac{\partial \bar{L}}{\partial y_2} = -2k_2 y_2 = \Delta y_2 \end{cases}$$

and

$$y = y + \Delta y = \begin{cases} y_1 + \Delta y_1 \\ y_2 + \Delta y_2 \end{cases}$$

Programming the pocket calculator typically involves the sequence of key strokes shown in Table 12-7.

Table 12-7 Gradient Optimization with Equality Constraint

Key Stroke Sequence		Comment
LBL-A		Set program step counter and pointer to begin at the first place in memory
R/S		Stop for data input
SRO-1		Store y_1 in register 1 (R-1)
R/S		
STO-2		Store y_2 in R-2
R/S	Input data	
STO-3		Store k_1 in R-3
R/S		
STO-4		Store k_2 in R-4
R/S		
STO-8		Store N in R-8
LBL-1		Label this step "1"
RCL-3		Recall R-3 (k_1)
1		
+		Add one to k_1
RCL-1		Recall y_1
f^{-1}		
$\sqrt{}$		Square y_1
×		$y_1^2 \times (1 + k_1)$
RCL-1		Recall y_1
4	Compute $\bar{L}$	
×		$4y_1$
−		$(1 - k_1)y_1^2 - 4y_1$
RCL-2		Recall y_2
f^{-1}		
$\sqrt{}$		y_2^2
+		$(1 + k_1)y_1^2 - 4y_1 + y_2^2$
4		
+		$(1 + k_1)y_1^2 - 4y_1 + y_2^2 + 4 = \bar{L}$
R/S		Display $(\bar{L})$
STO-6		Store $\bar{L}$ in R-6
RCL-3		Recall k_1
2		
×		$2k_1$
1		

Table 12-7 (*Continued*)

Key Stroke Sequence		Comment
+		$(1+2k_1)$
RCL-1	Compute	y_1
$\times$		$y_1(1+2k_1)$
2	$\Delta y_1 = \dfrac{-\partial \overline{L}}{\partial y_1} k_2$	
$-$		$(1+2k_1)y_1-2$
RCL-4		k_2
$\times$		$k_2[(1+2k_1)y_1-2]$
2		
$\times$		$2[(1+2k_1)y_1-2]k_2$
CHS		$-2[(1+2k_1)y_1-2]k_2 = \Delta Y_1$
R/S		Display ΔY_1
RCL-1		y_1
+	Compute	$y_1 + \Delta y_1 = y$
R/S	new y_1	Display y_1
STO-1		Store y_1 in R-1
RCL-2		Recall y_2
RCL-4		Recall k_2
$\times$	Compute	$k_2 y_2$
2	$\Delta y_2 = \dfrac{-\partial \overline{L}}{\partial y_2} k_2$	
$\times$		
CHS		$-2k_2 y_2 = \Delta y_2$
R/S		Display Δy_2
RCL-2		Recall y_2
+	Compute	$y_2 + \Delta y_2 = y_2$
R/S	new y_2	Display $(y_2 + \Delta y_2)$
STO-2		Store y_2 in R-2
DSZ	Return to top of	
GTO-1	program for iteration	
RCL-6	Display $\overline{L}$	
R/S		
RCL-1	Display y_1	
R/S		
RCL-2	Display y_2	
R/S		
RTN		

Let us take the example from Section 12-6 ($k_1=35$). Our first task is to find a value of k_2 that will ensure stability in the gradient search process. We see from Table 12-8 that $k_2=0.01$ can provide stable iterations at $k_1=35$ and is reasonably large to permit quick convergence (Table 12-8).

Table 12-8 Gradient Optimization Results for Various k_1

	k_1	k_2	$(\bar{L})_{10}$	$(y_1)_{10}$	$(y_2)_{10}$
Stable	0.0001	35	29.51091749	0.87049244	0.99800180
	0.001	35	6.55662021	0.23828719	0.98017904
	0.01	35	4.61180996	0.028335500	0.81707281
Unstable	0.05	35	$4.649658075 \times 10^{15}$	$6.932487456 \times 10^{7}$	0.34867844
	0.10	35	$5.032858806 \times 10^{21}$	$1.560738178 \times 10^{11}$	0.10737418

Using $k_2=0.01$, $k_1=35$, and $y_1=1=y_2$, we find after 100 steps that

$$(\bar{L})_{100}=3.93420285$$

$$(y_1)_{100}=0.02816901$$

$$(y_2)_{100}=0.13261956$$

and after 200 steps that

$$(\bar{L})_{200}=3.91621180$$

$$(y_1)_{200}=0.02816907$$

$$(y_2)_{200}=0.01758795$$

An approach that involves 200 steps but gives more confidence that the stationary point is in the neighborhood of the estimate made with the program is to make five 40-step searches for the stationary point and, as before, begin the searches from different quadrants as well as the "average" solution point. The value of k_1 is then selected so that in 40 steps the search will cover the region of the expected solution. In our case, we expect a solution in the neighborhood of $(0,0)$, so that 40 steps of 0.05 will cover the region from -1 to $+1$. Then the task is to find a reasonably high value of k_2 that will permit stable iterations. We see from Table 12-9 that a $k_2 = 7.5$ results in stable searches for the singular point. Then four 40-step searches provide the results given in Table 12-10. This approach results in more confidence that the stationary point for

$$L(y_1, y_2)$$

subject to the constraint $y_1 = 0$ is located at

$$y_1 = 0 \quad \text{(by the constraint)}$$

$$y_2 = 0 \quad \text{(by analysis)}$$

Table 12-9 Gradient Optimization Results for Various k_2

	Number of Iterations	k_2	k_1	$(\bar{L})_{10}$	$(y_1)_{10}$	$(y_2)_{10}$
Stable	10	1	0.05	2.35474379	0.67608251	0.34867844
	10	2.5	0.05	3.20535903	0.33340324	0.34867844
	10	5.0	0.05	3.62116902	0.18181818	0.34867844
	10	7.5	0.05	3.80010179	0.13029079	0.34867844
Unstable	10	10.0	0.05	57.99696974	2.44195747	0.34867844
	10	12.5	0.05	54447.09556	101.5703041	0.34867844

Table 12-10 Gradient Optimization using Penalty Function Method for Satisfying Equality Constraint—Average Result Technique

Initial Values				
y_1	y_2	$(\bar{L})_{40}$	$(y_1)_{40}$	$(y_2)_{40}$
1	1	3.63308223	0.12500000	0.01478088
-1	1	3.63308222	0.12500000	0.01478088
-1	-1	3.63308222	0.12500000	-0.01478088
$+1$	-1	3.63308223	0.12500000	-0.01478088

Average of all estimates of y_1 and y_2 0.12500000   0.00000000 or

$$y_1 = 0.125$$

(both by the analysis above)

$$y_2 = 0$$

We can make a test of the expected answer to complete the 200-step process as follows:

$$y_1 = 0.125$$

$$y_2 = 0.0$$

Then

$$(\bar{L})_{40} = 3.63281250$$

$$(y_1)_{40} = 0.12500000$$

$$(y_2)_{40} = 0.00000000$$

It is apparent that the stationary point of $L(y_1, y_2)$ is in the neighborhood of

$$\text{Stationary point for } \bar{L} \left\{ \begin{array}{ll} y_1 = 0.125, & y_1 = 0.0 \\[2ex] y_2 = 0.0, & y_2 = 0.0 \end{array} \right\} \text{Stationary point for } L$$

Of the two approaches to penalty function searches on the pocket calculator, the latter is recommended because more information on the local topology of $\bar{L}$ is used to generate the estimate of the values of y_1 and y_2 at the point where $\bar{L}$ (and thereby L) is a minimum.

12-7 REFERENCES

For this chapter consult A. E. Bryson and Y. C. Ho's excellent book, *Applied Optimal Control* (Blaisdell Publishing Company, 1969), Chapter 1. The examples used in this chapter were first presented by Bryson and Ho at their outstanding seminar on Applied Optimal Control.

APPENDIX 1

SOME TRICKS
OF THE POCKET
CALCULATOR TRADE

In the course of writing this book, a number of interesting "special methods" were offered by many colleagues. Unfortunately, the list is far longer than might be conveniently included in a single chapter. This appendix presents some of these methods, selected on the basis of their usefulness in pocket calculator analysis or because they are novel and interesting.

A1-1 π AND e ON THE FOUR-FUNCTION CALCULATOR

An easy-to-remember sequence of numbers, which will generate π with an error of only 4×10^{-7}, is

$$11 \ 33 \ 55$$

We see that this set of numbers is made up of double entries of the first three odd digits of the positive numbers. Then

$$\hat{\pi} = \frac{355}{113} = \pi + \epsilon$$

where

$$\epsilon \leqslant 4 \times 10^{-7}$$

The key stroke sequence for generating π on the four-function calculator is

$$(355)$$

$$\div$$

$$(113)$$

$$=$$

$$\boxed{3.1415929}$$

A similar ratio generating an approximation of the number e that was published by Texas Instruments Corporation in their applications guide is

$$\frac{193}{71} = 2.7183098$$

This ratio is not easily remembered, except perhaps by noticing that each digit is an odd number and that the digits appear in the sequence

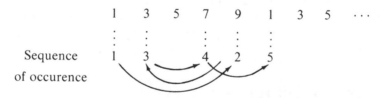

with the first three in the numerator and the last two in the denominator. The result of the ratio is accurate only to the fourth digit (i.e., 2.718), which is one digit fewer than must be remembered in the ratio (i.e., 1, 9, 3, 7, and 1). One might as well memorize e to five places:

$$e = 2.71828(1828\cdots)$$

The author devised a simpler, more accurate, and more easily remembered sequence of zeros and odd numbers (as used in the π sequence):

$$00\ 11\ 33\ 55\ 77\ 99$$

The procedure is as follows:

1. Cancel the 77 and 11 pairs (symmetric operation).
2. Put a decimal place after the first zero and a parentheses before the

last 9 (symmetric operation) to obtain

$$(0.0\ 33\ 55\ 9)9$$

This product is $e/9$ to six digits when rounded at the sixth digit. Thus

$$e = (0.0\ 33\ 55\ 9)9 \times 9|_{\text{rounded}}$$

In evaluating e in this manner, the key stroke sequence is

$$(0.0\ 33\ 55\ 9)$$

$$\times$$

$$(9)$$

$$=$$

$$=$$

$$\boxed{2.718279} \rightarrow 2.71828 \text{ when rounded}$$

The relative error in this evaluation of e is less than $7 \times 10^{-5}\%$.

To discover such sequences, the calculator becomes a research tool. When trying to find interesting ways to generate approximations to often-used numbers, one can begin by repeated calculator operations on a number and look for an interesting pattern. For example, four divisions of π by 6 on an eight-digit calculator will result in the number

$$0.002424$$

which is, curiously enough,

$$6 \times 4 \times 10^{-4} + 6 \times 4 \times 10^{-6}$$

Thus

$$\pi \cong 6^4 (6 \times 4 \times 10^{-4} + 6 \times 4 \times 10^{-6}) \equiv 3.141504$$

for a relative error of

$$\frac{0.0000886}{03.1415926} \times 100 = 0.00282\%$$

Note that only the two even numbers, 4 and 6, are used in this evaluation of π. This particular approximation was worked out at the time of this writing as an illustration of the interesting properties of number approximations that can be found with the aid of a simple eight-digit calculator.

A1-2 TRUNCATING A NUMBER ON A CALCULATOR THAT DOES NOT HAVE EXPONENTIAL NOTATION

Truncating a number is the process of reducing a number made up of an integer and fractional parts to an integer. For example, when the numbers

$$1.21743$$

$$247.41715$$

$$5764.88177$$

are truncated, they result in the numbers

$$1.0$$

$$247.0$$

$$5764.0$$

Notice that there is no rounding. An interesting approach to truncating any number in the registers of a pocket calculator *that does not have* scientific notation is the following:

  1. Divide the number by $1000\cdots$, where the number of zeros fills the rest of the display register.
  2. Multiply the result of step (1) by the divisor in step (1)—$1000\cdots$.

The result is the truncated number being sought. The key stroke sequence is

(number in display register)

$\div$

$(100000\cdots)$

$\times$

$(10000\cdots)$

$=$

truncated number in

display register

A1-3 LUKASIEWIC'S ALGORITHM FOR EVALUATING ANY FUNCTION ON A MACHINE WITH REVERSE-POLISH NOTATION PLUS AN OPERATIONAL STACK

Step 1 Write function in serial form.

Step 2 Key in first number.

Step 3 Compute all functions of the single number and enter them in the stack (keyboard operations such as ln(x), 10^x, and sin(x).

Step 4 Compute all 2 number functions and enter them in the stack (keyboard operations such as $+$, $-$, $\times$, $\div$, x^y, and xy,).

Step 5 Key in next number, then repeat steps 3 through 5 until the function is evaluated.

This algorithm is flow-charted as shown in Figure A-1. Clearly, this algorithm requires an infinite number of registers in the stack to evaluate any function. The lower limit is two registers and a reasonable size is three for most commonly encountered scientific functions. Hewlett-Packard's HP-35, HP-45, and HP-65 calculators (the most popular Reverse-Polish plus stacks machines) all have four registers in their operational stack.

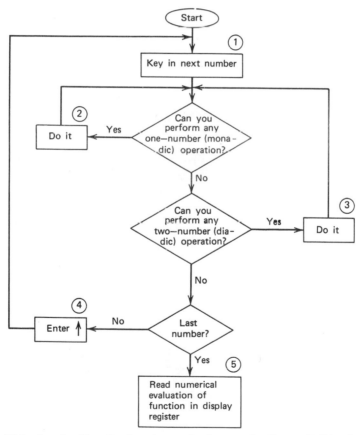

Figure A1-1 An algorithm for function evaluation on the Reverse-Polish plus stacks machines.

Example. Consider the expression

$$(A + B)\left[\ln(C + D)^{1/2} + E\right]$$

This complex expression can be evaluated according to the algorithm with the following key strokes on the HP-35 (circled number correlates evaluation to the algorithm flow chart operation):

key stroke sequence	
(A)	①
↑	④
(B)	①
+	③
↑	④
(C)	①
↑	④
(D)	①
+	③
√	②
ln	②
(E)	①
+	③
×	③

Result in
Display Register

A1-4 QUICK POLYNOMIAL APPROXIMATIONS FOR ANALYTIC SUBSTITUTION

A simple, not too accurate, but fast, procedure for developing polynomials that can be used to approximate functions is the following:

1. Identify a number of rational number conditions, $x_1, x_2, \ldots, x_3$, under which the function takes on rational number values, $c_1, c_2, \ldots, c_n$.

2. Prepare a polynomial with coefficients $a_1, a_2, \ldots, a_n$) which is evaluated using simultaneous equations.

Example. Approximate $\sin\theta$ with a second-order polynomial on the interval $0<\theta<90°$.

θ Degrees	$\sin\theta$
0	0
30	$\frac{1}{2}$
90	1

We use a quadratic equation $\sin\theta \approx a_1 + a_2\theta + a_3\theta^2$ to approximate $\sin\theta$ on the interval 0–90°. The coefficients are determined using the simultaneous equations

$$0 = a_1 + a_2(0) + a_3(0)^2$$

$$\tfrac{1}{2} = a_1 + a_2(30) + a_3(30)^2$$

$$1 = a_1 + a_2(90) + a_3(90)^2$$

By inspection we see that

$$a_1 = 0$$

This system of equations reduces to

$$\tfrac{1}{2} = a_2(30) + a_3(30)^2$$
$$1 = a_2(90) + a_3(90)^2$$

$$-\tfrac{3}{2} = a_2(90) - 3a_3(30)^2$$
$$1 = a_2(90) + a_3(90)^2$$

Summing we find:

$$(1 - \tfrac{3}{2}) = -\tfrac{1}{2} = a_3\left[(90)^2 - 3(30)^2\right]$$

$$a_3 = \frac{-\tfrac{1}{2}}{81\times 10^2 - 27\times 10^2} = \frac{-\tfrac{1}{2}}{54\times 10^2} = \frac{-1}{108\times 10^2}$$

$$\therefore a_3 = \frac{-1}{10800}$$

We can now use a_3 to compute a_2 by way of one of the simultaneous equations:

$$\tfrac{1}{2} = a_2(30) - \frac{900x}{10800}$$

$$a_2 = \frac{1}{30}\left(\frac{1}{2} + \frac{900}{10800}\right) = \frac{5400 + 900}{324000} = \frac{6300}{324000} = \frac{63}{3240}$$

The approximating polynomial then is

$$\sin\theta = \frac{63\theta}{3240} - \frac{\theta^2}{10800} + \epsilon \qquad (\theta \text{ in degrees})$$

for $0 \leqslant \theta \leqslant 90°$, where

$$100(\epsilon/\sin\theta) \leqslant 11.14\%$$

The characteristics of this approximation are seen given in Table A1-1.

Table A1-1

θ (degrees)	$\sin\theta$	$\left(\dfrac{63\theta}{3240} - \dfrac{\theta^2}{10800}\right)$	Relative Error Percent	Maclaurin Expansion $\left(\theta - \dfrac{\theta^3}{6}\right)$
0	.0	.0	$\lim\limits_{\theta \to 0}\left(\dfrac{\text{est }\sin\theta}{\sin\theta}\right) = -11.41$	
5	0.08715574	0.09490741	-8.89	0.08715570
10	0.17364818	0.18518519	-6.64	0.17364683
15	0.25881905	0.27083333	-4.64	0.25880881
20	0.34202015	0.35185185	-2.87	0.34197708
25	0.42261826	0.42824074	-1.33	0.42248706
30	0.5	0.5	0.00	0.49967418
35	0.57357644	0.56712963	$+1.24$	0.57287387
40	0.64278761	0.62962963	$+2.05$	0.64142155
45	0.70710678	0.68750000	$+2.77$	0.70465265
55	0.76604444	0.74074074	$+3.30$	0.81250684
60	0.86602540	0.83333337	$+3.77$	0.85580078
65	0.90630779	0.87268519	$+3.71$	0.89111986
70	0.93969262	0.90740741	$+3.44$	0.91779950
75	0.96592583	0.93750000	$+2.94$	0.93517512
80	0.98480775	0.96396296	$+2.22$	0.94258217
85	0.99619470	0.98379630	$+1.24$	0.93935606
90	1.0	1.0	0.0	0.92483223

The table showing a relative error as large as 11.41% should be enough to point up the inaccuracy associated with this method. It is worth mentioning, however, because the method can be useful for quick curve getting through experimental data or measurements that are known only to a few percent.

A1-5 A METHOD FOR COMPUTING RECIPROCALS ON THE FOUR-FUNCTION CALCULATOR

A straightforward, but often overlooked (even by manufacturers in their applications manuals) technique for evaluating the reciprocal of a number is to enter the number into the display and constant registers (usually done automatically for the constant register), divide the number by itself to enter 1 in the display register, and then divide again to find the reciprocal. The key stroke sequence is shown in Table A1-2.

Table A1-2

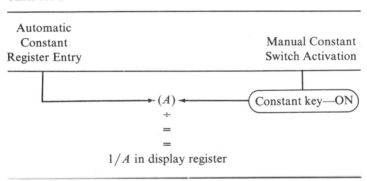

$1/A$ in display register

A1-6 ALPHA-NUMERICS ON THE POCKET CALCULATOR

Calculators are, by definition, capable solely of numerical manipulations and displays. An interesting aspect of our arabic-based alphabet and number system is that they have many common symbol shapes. Because of this there are words that can be spelled out using numbers. For example, in most calculator displays

$$\boxminus \square \mathrel{\raise.1ex\hbox{S}} \mathrel{\raise.1ex\hbox{S}} = 8055$$

Even more interesting is the fact that some letters are made up of upside-down numbers. A typical example is

$$\exists \text{ upside down} = E$$

The numbers displayed in the calculator display window that correspond to letters are shown in Table A1-3.

Table A1-3

Numeric Display Window	Corresponding Right-side up Letters	Corresponding Upsidedown Letters
0	0	0
1	*I* or *l*	*I* or *l*
2		
3		*E*
4	*y*	*h*
5	*S*	*S*
6	*G*	
7		*L*
8	*B*	*B*
9		*G*
Total	7	9

Note that there are three vowels among the upside-down letters but only two among the right-side-up letters. Also, there is one more consonant among the upside-down letters than among the right-side up letters. The author conjectures that it is for this reason that the upside-down set is more popular with pocket calculator innovators.

Examples of the better known pocket calculator "scrabble" words are the following (all to be viewed upside down):

Greeting	07734
Object	38079
Proper name	318808
Adjective	35007
Expletive	$57738.57734 \times 10^{40}$

MATRIX ANALYSIS
ON THE POCKET CALCULATOR

Matrix manipulations on the pocket calculator are fairly straightforward compared with matrix calculations for general-purpose computing. The matrices that the pocket calculator, and even the programmable pocket calculator, can operate on are small, hence can be easily manipulated manually if problems of ill-conditioned matrices are encountered. Here we concern ourselves with basic matrix operations for 2×2 and 3×3 matrices.

The most fundamental matrix operations are those of addition, subtraction, and multiplication of two matrices. Consider the two matrices

$$\mathbf{A} = \begin{bmatrix} a_1 & a_2 \\ a_3 & a_4 \end{bmatrix}, \qquad \mathbf{B} = \begin{bmatrix} b_1 & b_2 \\ b_3 & b_4 \end{bmatrix}$$

The sum of these two matrices is then given by

$$\mathbf{A} + \mathbf{B} = \begin{bmatrix} a_1 + b_1 & a_2 + b_2 \\ a_3 + b_3 & a_4 + b_4 \end{bmatrix}$$

and the difference by

$$\mathbf{A} - \mathbf{B} = \begin{bmatrix} a_1 - b_1 & a_2 - b_2 \\ a_3 - b_3 & a_4 - b_4 \end{bmatrix}$$

and the product by

$$AB = \begin{bmatrix} a_1b_1 + a_2b_3 & a_1b_2 + a_2b_4 \\ a_3b_1 + a_4b_3 & a_3b_2 + a_4b_4 \end{bmatrix}$$

The inverse of the 2×2 matrix **A** is given by

$$A^{-1} = \begin{bmatrix} \alpha_1 & \alpha_2 \\ \alpha_3 & \alpha_4 \end{bmatrix}$$

where

$$\alpha_1 = a_4/\det$$

$$\alpha_2 = a_3/\det$$

$$\alpha_3 = a_2/\det$$

$$\alpha_4 = a_1/\det$$

The determinate can be numerically evaluated as

$$\det A = a_1a_4 - a_2a_3$$

In a similar fashion, the 3×3 matrix operations can be defined for the sum, difference, and product. However, the inverse of a 3×3 matrix is defined somewhat differently. Consider now the matrix **A** defined by

$$A = \begin{bmatrix} a_1 & b_1 & c_1 \\ a_2 & b_2 & c_2 \\ a_3 & b_3 & c_2 \end{bmatrix}$$

which has the inverse

$$A^{-1} = \begin{bmatrix} \alpha_1 & \alpha_4 & \alpha_7 \\ \alpha_2 & \alpha_5 & \alpha_8 \\ \alpha_3 & \alpha_6 & \alpha_9 \end{bmatrix}$$

where the alphas can be numerically evaluated with the equations

$$\alpha_1 = (b_2 c_3 - b_3 c_2)/\det$$

$$\alpha_2 = (a_3 c_2 - a_2 c_3)/\det$$

$$\alpha_3 = (a_2 b_3 - a_3 b_2)/\det$$

$$\alpha_4 = (b_3 c_1 - b_1 c_3)/\det$$

$$\alpha_5 = (a_1 c_3 - a_3 c_1)/\det$$

$$\alpha_6 = (a_3 b_1 - a_1 b_3)/\det$$

$$\alpha_7 = (b_1 c_2 - b_2 c_1)/\det$$

$$\alpha_8 = (a_2 c_1 - a_1 c_2)/\det$$

$$\alpha_9 = (a_1 b_2 - a_2 b_1)/\det$$

Here the determinate can be numerically evaluated with the equation

$$\det = a_1 b_2 c_3 + a_2 b_3 c_1 + a_3 b_1 c_2 - a_3 b_2 c_1 - a_2 b_1 c_3 - a_1 b_3 c_2$$

Clearly the matrix inversion will work only if the determinate is nonzero.

Another important matrix manipulation that is frequently encountered and can be easily evaluated on the pocket calculator is the determination of the characteristic equation for the matrix A. That is,

$$A - \lambda I = -\lambda^3 + d_1 \lambda^2 + d_2 \lambda + d_3 = 0$$

Here d_1 through d_3 can be given by the equation

$$d_1 = a_1 + b_2 + c_3$$

$$d_2 = a_3 c_1 + a_2 b_1 + b_3 c_2 - a_1 b_2 - a_1 c_3 - b_2 c_3$$

$$d_3 = \det = a_1 b_2 c_3 + a_2 b_3 c_1 + a_3 b_1 c_2 - a_3 b_2 c_1$$

$$- a_2 b_1 c_3 - a_1 b_3 c_2$$

Using these equations, the matrix manipulations associated with many vector-matrix operations can be determined for second- and third-order matrices. Beyond the second- and third-order matrix analysis, the evaluation on the pocket calculator, though possible, becomes somewhat tedious.

COMPLEX NUMBERS
AND FUNCTIONS

While it is not the purpose of this book to teach, complex variable theory, it will be discussed to review the concepts in complex numbers and functions that are pertinent to the evaluation of advanced mathematical functions. Complex numbers written in Cartesian form are as follows:

$$z = x + iy$$

Complex numbers in Cartesian form can be written in polar form:

$$z = re^{i\theta} = r(\cos\theta + i\sin\theta)$$

It is apparent that the modulus of the complex number is

$$|z| = (x^2 + y^2)^{1/2} = r$$

Similarly, the argument of a complex number is given by

$$\arg(z) = \tan^{-1}\left(\frac{y}{x}\right) = \theta$$

It is common to refer to the real and imaginary parts of a complex number:

$$\text{Re}(z) = R(z) = x = r\cos\theta = \text{real part}$$

$$\text{Im}(z) = I(z) = y = r\sin\theta = \text{imaginary part}$$

The complex conjugate of a complex variable c is given by

$$\bar{z} = x - iy = z_c$$

It is clear from the definition of the modulus of a complex number that

$$|z_c| = |z|$$

Similarly the equation for the argument of a complex number allows

$$\arg(z_c) = -\arg(z)$$

It is worth remembering that the complex conjugate is used in clearing the complex form of the denominator of a complex number.

The multiplication and division of two complex numbers

$$z_1 = x_1 + iy_1 \quad \text{and} \quad z_2 = x_2 + iy_2$$

are given by

$$z_1 z_2 = x_1 x_2 - y_1 y_2 + i(x_1 y_2 + x_2 y_1)$$

and

$$\frac{z_1}{z_2} = \frac{z_1 z_{2_c}}{|z_2|^2} = \frac{x_1 x_2 + y_1 y_2 + i(x_2 y_1 - x_1 y_2)}{x_2^2 + y_2^2}$$

It is apparent that

$$|z_1 z_2| = |z_1||z_2|$$

Similarly we see that

$$\arg(z_1 z_2) = \arg z_1 + \arg z_2$$

$$\left|\frac{z_1}{z_2}\right| = \frac{|z_1|}{|z_2|}$$

$$\arg\left(\frac{z_1}{z_2}\right) = \arg(z_1) - \arg(z_2)$$

Powers of complex numbers can be written in polar form:

$$z^n = r^n e^{in\theta}$$

which is equivalent to

$$z^n = r^n \cos n\theta + i r^n \sin n\theta, \qquad (n = 0, \pm 1, \pm 2, \dots)$$

In particular,

$$z^2 = x^2 - y^2 + i(2xy)$$

$$z^3 = x^3 - 3xy^2 + i(3x^2y - y^3)$$

$$z^4 = x^4 - 6x^2y^2 + y^4 + i(4x^3y - 4xy^3)$$

$$z^5 = x^5 - 10x^3y^2 + 5xy^4 + i(5x^4y - 10x^2y^3 + y^5)$$

In general, we can write

$$z^n = \left[x^n - \binom{n}{2} x^{n-2}y^2 + \binom{n}{4} x^{n-4}y^4 - \cdots \right]$$

$$+ i \left[\binom{n}{2} x^{n-1}y - \binom{n}{3} x^{n-3}y^3 + \cdots \right], \qquad (n = 1, 2, \ldots)$$

Furthermore, if the nth power of z is written in the form

$$z^n = u_n + iv_n$$

then

$$z^{n+1} = u_{n+1} + iv_{n+1}$$

where

$$u_{n+1} = xu_n - yv_n$$

$$v_{n+1} = xv_n + yu_n$$

For negative powers of a complex number,

$$\frac{1}{z} = \frac{z_c}{|z|^2} = \frac{x - iy}{x^2 + y^2}$$

and more generally

$$\frac{1}{z^n} = \frac{z_c^n}{|z|^{2n}} = (z^{-1})^n$$

The roots of a complex number are easily interpreted in polar form:

$$z^{1/2} = r^{1/2}e^{i\theta/2} = r^{1/2}\cos\left(\frac{\theta}{2}\right) + ir^{1/2}\sin\left(\frac{\theta}{2}\right)$$

If data are greater than $-\pi$ but smaller than or equal to $+\pi$, equation computes the principal root. The other root has the opposite sign, of course. The principal root is always given by

$$z^{1/2}=\left[\tfrac{1}{2}(r+x)\right]^{1/2}\pm i\left[\tfrac{1}{2}(r-x)\right]^{1/2}=u\pm iv$$

where $2uv=y$ and the ambiguous sign is taken to be the same as the sign of y.

In general, then, the nth root of the complex number z is given in polar form by

$$z^{1/n}=r^{1/n}e^{i\theta/n}$$

Again, this equation computes the principal root if θ is greater than $-\pi$, but smaller than or equal to $+\pi$. The other roots are computed from the expression

$$r^{1/n}e^{i(\theta+2\pi k)/n},\qquad (k=1,2,3,\ldots,n-1)$$

KEY STROKE SEQUENCES FOR COMPLEX VARIABLE ANALYSIS AND HYPERBOLIC, INVERSE HYPERBOLIC FUNCTIONS

The Hewlett-Packard Company has graciously given permission to publish certain pages from their HP-35 MATH PAC that will help the reader perform complex variable analysis on the pocket calculator. Key stroke sequences for commonly encountered complex variable analysis and certain hyperbolic functions are given. They are useful for analysis on the HP-35. They are also useful for showing the details of the mathematics required to do the analysis on any calculator. The Hewlett-Packard corporation has à superb pocket calculator product support organization that publishes handbooks of keystroke sequences for solving problems in many fields (business, real-estate, science, etc.). The HP-35 MATH PAC handbook of keystroke sequences can be obtained from any Hewlett-Packard representative or from the Hewlett-Packard Company, Advanced Products Division, 10900 Wolfe Road, Cupertino, California 95014.

Complex Number Operations

Complex add

Formula:

$$(a_1 + ib_1) + (a_2 + ib_2) = (a_1 + a_2) + i(b_1 + b_2)$$

$$= u + iv$$

Example:

$$(3 + 4i) + (7.4 - 5.6i) = 10.4 - 1.6i$$

LINE	DATA	OPERATIONS						DISPLAY	REMARKS
1	a_1	↑							
2	a_2	+						u	
3	b_1	↑							
4	b_2	+						v	

Complex subtract

Formula:

$$(a_1 + ib_1) - (a_2 + ib_2) = (a_1 - a_2) + i(b_1 - b_2)$$

$$= u + iv$$

Example:

$$(3 + 4i) - (7.4 - 5.6i) = -4.4 + 9.6i$$

LINE	DATA	OPERATIONS						DISPLAY	REMARKS
1	a_1	↑							
2	a_2	−						u	
3	b_1	↑							
4	b_2	−						v	

Complex multiply

(Use for short data input)

Formula:

$$(a_1 + ib_1)(a_2 + ib_2) = (a_1 a_2 - b_1 b_2) + i(a_1 b_2 + a_2 b_1)$$

$$= u + iv$$

Example:

$$(3 + 4i)\,(7 - 2i) = 29 + 22i$$

LINE	DATA	OPERATIONS					DISPLAY	REMARKS
1	b_1	↑						
2	a_1	↑	↑					
3	a_2	×						
4	b_2	R↓	R↓	R↓	STO	×		
5		−					u	
6	a_2	RCL	×	R↓	R↓	R↓		
7	b_2	×	+				v	

Complex multiply, alternate method

(Data are entered only once, use for long data input)

Method:

If a_1, a_2, b_1, b_2 are all non-zero, and if $\theta = \theta_1 + \theta_2 \neq 90°$ or $-90°$, then

$$(a_1 + ib_1)\,(a_2 + ib_2) = \frac{b_1 \cdot b_2}{\sin\theta_1 \cdot \sin\theta_2}\,(\cos\theta + i\sin\theta) = u + iv.$$

Note: $a_1 + ib_1 = r_1 e^{i\theta_1}$,

$a_2 + ib_2 = r_2 e^{i\theta_2}$

Example:

$$(3 + 4i)\,(5 - 12i) = 63 - 16i$$

LINE	DATA	OPERATIONS					DISPLAY	REMARKS
1	a_1	↑						
2	b_1	STO	x⇄y	÷	arc	tan		
3	a_2	↑						
4	b_2	↑	R↓	x⇄y	÷	arc		
5		tan	RCL	R↓	R↓	R↓		
6		×	STO	R↓	↑	R↓		
7		+					D	If D = 90 or −90, stop.
								Use other method.
8		RCL	R↓	STO	R↓	sin		
9		x⇄y	sin	×	÷	RCL		
10		cos	×	↑	↑	RCL		
11		tan	×	x⇄y			u	
12		x⇄y					v	

342

Complex divide

(Use for short data input)

Formula:

$$(a_1 + ib_1) \div (a_2 + ib_2) = \frac{(a_1 a_2 + b_1 b_2) + i(a_2 b_1 - a_1 b_2)}{a_2{}^2 + b_2{}^2}$$

$$= u + iv$$

Example:

$$\frac{3 + 4i}{7 - 2i} = .245 + .64i$$

LINE	DATA	OPERATIONS					DISPLAY	REMARKS
1	b_1	↑						
2	a_1	↑	↑					
3	a_2	×						
4	b_2	R↓	R↓	R↓	STO	×		
5		+						
6	a_2	RCL	×	R↓	R↓	R↓		
7	b_2	×	−	STO				
8	a_2	↑	×					
9	b_2	↑	×	+	RCL	x⇄y		
10		STO	÷	x⇄y	RCL	÷	u	
11		x⇄y					v	

Complex divide, alternate method

(Data are entered only once; use for long data input)

Formula:

If a_1, a_2, b_1, b_2 are all non-zero, and if $\theta = \theta_1 - \theta_2 \neq 90°$ or $-90°$, then

$$(a_1 + ib_1) \div (a_2 + ib_2) = \left(\frac{b_1}{b_2} \Big/ \frac{\sin \theta_1}{\sin \theta_2} \right) (\cos \theta + i \sin \theta) = u + iv$$

Note: $a_1 + ib_1 = r_1 e^{i\theta_1}$,

$a_2 + ib_2 = r_2 e^{i\theta_2}$.

Example:

$$(63 - 16i) \div (5 - 12i) = 3 + 4i$$

LINE	DATA	OPERATIONS					DISPLAY	REMARKS
1	a_1	↑						
2	b_1	STO	x⇄y	÷	arc	tan		
3	a_2	↑						
4	b_2	↑	R↓	x⇄y	÷	arc		
5		tan	RCL	R↓	R↓	R↓		
6		÷	STO	R↓	↑	R↓		
7		−					D	If D = 90 or −90, stop.
								Use other method.
8		RCL	R↓	STO	R↓	sin		
9		x⇄y	sin	÷	÷	RCL		
10		cos	x	↑	↑	RCL		
11		tan	x	x⇄y			u	
12		x⇄y					v	

Complex reciprocal

Formula:

$$\frac{1}{a + ib} = \frac{a - ib}{a^2 + b^2} = u + iv$$

Example:

$$\frac{1}{2 + 3i} = .15 - .23i$$

LINE	DATA	OPERATIONS					DISPLAY	REMARKS
1	b	↑						
2	a	↑	↑	x	STO	CLX		
3		+	R↓	x	RCL	+		
4		STO	÷	CHS	x⇄y	RCL		
5		÷					u	
6		x⇄y					v	

344

Complex absolute value

Formula:

$$|a + ib| = \sqrt{a^2 + b^2}$$

Example:

$$|3 + 4i| = 5$$

LINE	DATA	OPERATIONS					DISPLAY	REMARKS
1	a	↑	x					
2	b	↑	x	+	√x̄			

Complex square

Formula:

$$(a + ib)^2 = (a^2 - b^2) + i(2ab) = u + iv$$

Example:

$$(7 - 2i)^2 = 45 - 28i$$

LINE	DATA	OPERATIONS					DISPLAY	REMARKS
1	a	↑	↑	x				
2	b	STO	↑	x	−		u	
3		x⇄y	RCL	x	2	x	v	

Complex square root

Formula:

$$\sqrt{a + ib} = \pm \left[\sqrt{\frac{a + \sqrt{a^2 + b^2}}{2}} + i\, \frac{b}{2\sqrt{\dfrac{a + \sqrt{a^2 + b^2}}{2}}} \right] = \pm(u + iv)$$

Example:

$$\sqrt{7 + 6i} = \pm(2.85 + 1.05i)$$

LINE	DATA	OPERATIONS					DISPLAY	REMARKS
1	b	↑						
2	a	STO	x⇄y					If b = 0 and a < 0, go to 7;
								if b = 0 and a ≥ 0, go to 9.
3		↑	↑	x	RCL	↑		
4		x	+	√x	RCL	+		
5		2	÷	√x	STO		u	
6		2	RCL	x	÷		v	Stop
7		RCL	CHS	√x	x⇄y		u	
8		x⇄y					v	Stop
9		RCL	√x				u	
10		x⇄y					v	

Complex natural logarithm (base e)

Formula:

$$\ln (a + ib) = \ln \left(\sqrt{a^2 + b^2}\right) + i\left(\text{arc tan } \frac{b}{a}\right)\frac{\pi}{180}$$

$$= u + iv$$

Example:

$$\ln i = 1.57i$$

LINE	DATA	OPERATIONS					DISPLAY	REMARKS
1	b	↑						
2	a							If a ≠ 0, go to 4
3		CLX	EEX	CHS	9	9		
4		STO	x⇄y	↑	↑	x		
5		RCL	↑	x	+	√x		
6		ln					u	
7		x⇄y	RCL	÷	arc	tan		If a > 0, go to 9
8		1	8	0	+			
9		π	x	1	8	0		
10		÷					v	

Complex exponential

Formula:

$$e^{(a+ib)} = e^a (\cos \theta + i \sin \theta) = u + iv$$

where

$$\theta = \frac{180b}{\pi}$$

Example:

$$e^{1.570796327i} = i$$

LINE	DATA	OPERATIONS					DISPLAY	REMARKS
1	b	↑						
2	a	e^x	STO	x⇄y	1	8		
3		0	×	π	÷	↑		
4		cos	RCL	×			u	
5		x⇄y	sin	RCL	×		v	

Complex exponential (t^{a+ib})

Formula:

$$t^{a+ib} = e^{(a+ib)\ln t} = u + iv$$

Restriction:

$$(t > 0)$$

Example:

$$2^{3+4i} = -7.46 + 2.89i$$

LINE	DATA	OPERATIONS					DISPLAY	REMARKS
1	b	↑						
2	a	x⇄y						
3	t	ln	STO	×	x⇄y	RCL		
4		×	e^x	STO	x⇄y	1		
5		8	0	×	π	÷		
6		↑	cos	RCL	×		u	
7		x⇄y	sin	RCL	×		v	

Complex number to integral power

Formula:

$$(a + ib)^n = r^n (\cos n\theta + i \sin n\theta) = u + iv$$

where: $r = \sqrt{a^2 + b^2}$,

$\theta = \text{arc tan } \dfrac{b}{a}$, and

n is a positive integer.

Example:

$$(3 + 4.5i)^5 = 926.44 - 4533.47i$$

LINE	DATA	OPERATIONS					DISPLAY	REMARKS
1	b	↑						
2	a							If a ≠ 0, go to 4
3		CLX	EEX	CHS	9	9		
4		STO	x⇄y	↑	↑	x		
5		RCL	↑	x	+	√x̄		
6		x⇄y	RCL	÷	arc	tan		If a ⩾ 0, go to 8
7		1	8	0	+			
8		STO						
9	n	↑	↑	R↓	R↓	R↓		
10		xʸ	RCL	x⇄y	STO	CLX		
11		+	x	↑	cos	RCL		
12		x					u	
13		x⇄y	sin	RCL	x		v	

348

Integral roots of complex number

Formula:

$$(a + ib)^{\frac{1}{n}} = r^{\frac{1}{n}}\left(\cos\frac{\theta + 360k}{n} + i\sin\frac{\theta + 360k}{n}\right)$$

$$= u_k + iv_k$$

where: n is a positive integer, and

k = 0, 1, ..., n − 1.

Example:

5 + 3i has three cube roots:

$$u_0 + iv_0 = 1.77 + .32i$$
$$u_1 + iv_1 = -1.16 + 1.37i$$
$$u_2 + iv_2 = -.61 - 1.69i$$

LINE	DATA	OPERATIONS						DISPLAY	REMARKS
1	b	↑							
2	a								If a ≠ 0, go to 4
3		CLX	EEX	CHS	9	9			
4		STO	x⇄y	↑	↑	x			
5		RCL	↑	x	+	√x			
6		x⇄y	RCL	÷	arc	tan			If a ≥ 0, go to 8
7		1	8	0	+				
8		x⇄y							
9	n	↑	¹/ₓ	R↓	R↓	x⇄y			
10		R↓	xʸ	STO	CLX	+			
11		÷	↑	↑	cos	RCL			
12		x						u₀	
13		x⇄y	sin	RCL	x			v₀	
14		R↓	R↓						Perform lines 14–18 for k = 1,
									2, ..., n−1
15		3	6	0	↑				
16	n	÷	+	↑	↑	cos			
17		RCL	x					uₖ	
18		x⇄y	sin	RCL	x			vₖ	

Complex number to a complex power

Formula:

$$\text{If } a_1 + ib_1 \neq 0,$$

$$(a_1 + ib_1)^{(a_2 + ib_2)} = e^{(a_2 + ib_2)\ln(a_1 + ib_1)} = u + iv$$

Example:

$$(1 + i)^{(2-i)} = 1.49 + 4.13i$$

LINE	DATA	OPERATIONS						DISPLAY	REMARKS
1	b_1	↑							
2	a_1								If $a_1 \neq 0$, go to 4
3		CLX	EEX	CHS	9	9			
4		STO	x⇄y	↑	↑	x			
5		RCL	↑	x	+	√x			
6		ln	x⇄y	RCL	÷	arc			
7		tan							If $a_1 \geqslant 0$, go to 9
8		1	8	0	+				
9		π	x	1	8	0			
10		÷	x⇄y	↑	↑				
11	a_2	x							
12	b_2	R↓	R↓	R↓	STO	x			
13		−							
14	a_2	RCL	x	R↓	R↓	R↓			
15	b_2	x	+	x⇄y	e^x	STO			
16		x⇄y	1	8	0	x			
17		π	÷	↑	cos	RCL			
18		x						u	
19		x⇄y	sin	RCL	x			v	

Complex root of a complex number

Formula:

If $a_1 + ib_1 \neq 0$

$$(a_1 + ib_1)^{\frac{1}{a_2 + ib_2}} = e^{[\ln(a_1 + ib_1)]/(a_2 + ib_2)}$$

$$= u + iv$$

Example:

Find the $(2 - i)^{th}$ root of $1.49 + 4.126i$.

Answer: $1 + i$

LINE	DATA	OPERATIONS						DISPLAY	REMARKS
1	b_1	↑							
2	a_1								If $a_1 \neq 0$, go to 4
3		CLX	EEX	CHS	9	9			
4		STO	x⇌y	↑	↑	x			
5		RCL	↑	x	+	√x			
6		ln	x⇌y	RCL	÷	arc			
7		tan							If $a_1 > 0$, go to 9
8		1	8	0	+				
9		π	x	1	8	0			
10		÷	x⇌y	↑	↑				
11	a_2	x							
12	b_2	R↓	R↓	R↓	STO	x			
13		+							
14	a_2	RCL	x	R↓	R↓	R↓			
15	b_2	x	−	STO					
16	a_2	↑	x						
17	b_2	↑	x	+	RCL	x⇌y			
18		STO	÷	x⇌y	RCL	÷			
19		eˣ	STO	x⇌y	1	8			
20		0	x	π	÷	↑			
21		cos	RCL	x				u	
22		x⇌y	sin	RCL	x			v	

Logarithm of a complex number to a complex base

Formula: $$\log_{(a_1 + ib_1)}(a_2 + ib_2) = \frac{\ln(a_2 + ib_2)}{\ln(a_1 + ib_1)} = u + iv$$

351

$$a_1 + ib_1 \neq 0$$

Example: $\log_{(1+i)} (1.49 + 4.126i) = 2 - i$

$$(a_3 = .34657359, \quad b_3 = .7853981633)$$

LINE	DATA	OPERATIONS					DISPLAY	REMARKS
1	b_1	↑						
2	a_1							If $a_1 \neq 0$, go to 4
3		CLX	EEX	CHS	9	9		
4		STO	x⇄y	↑	↑	x		
5		RCL	↑	x	+	√x̄		
6		ln					a_3	
7		x⇄y	RCL	÷	arc	tan		If $a_1 > 0$, go to 9
8		1	8	0	+			
9		π	x					
10		1	8	0	÷		b_3	
11		CLR						
12	b_2	↑						
13	a_2							If $a_2 \neq 0$, go to 15
14		CLX	EEX	CHS	9	9		
15		STO	x⇄y	↑	↑	x		
16		RCL	↑	x	+	√x̄		
17		ln	x⇄y	RCL	÷	arc		
18		tan						If $a_2 > 0$, go to 20
19		1	8	0	+			
20		π	x	1	8	0		
21		÷	x⇄y	↑	↑			
22	a_3	x						
23	b_3	R↓	R↓	R↓	STO	x		
24		+						
25	a_3	RCL	x	R↓	R↓	R↓		
26	b_3	x	−	STO				
27	a_3	↑	x					
28	b_3	↑	x	+	RCL	x⇄y		
29		STO	÷	x⇄y	RCL	÷	u	
30		x⇄y						v	

Complex Trigonometric and Hyperbolic Functions

In this section all angles in the equations are in radians.

Complex sine

Formula:

$$\sin(a + ib) = \sin a \cosh b + i \cos a \sinh b$$

$$= u + iv$$

Example:

$$\sin(2 + 3i) = 9.154 - 4.1689i$$

LINE	DATA	OPERATIONS					DISPLAY	REMARKS
1	b	↑						
2	a	↑	1	8	0	x		
3		π	÷	STO	sin	x⇄y		
4		↑	↑	R↓	R↓	eˣ		
5		↑	¹/ₓ	+	2	÷		
6		x	RCL	x⇄y	STO		u	
7		CLX	+	cos	x⇄y	eˣ		
8		↑	¹/ₓ	−	2	÷		
9		x					v	

Complex cosine

Formula:

$$\cos(a + ib) = \cos a \cosh b - i \sin a \sinh b$$

$$= u + iv$$

Example:

$$\cos(2 + ei) = -4.189 - 9.109i$$

LINE	DATA	OPERATIONS					DISPLAY	REMARKS
1	b	↑						
2	a	↑	1	8	0	x		
3		π	÷	STO	cos	x⇄y		
4		↑	↑	R↓	R↓	eˣ		
5		↑	¹/ₓ	+	2	÷		
6		x	RCL	x⇄y	STO		u	
7		CLX	+	sin	x⇄y	eˣ		
8		↑	¹/ₓ	–	2	÷		
9		x	CHS				v	

Complex tangent

Formula:

$$\tan (a + ib) = \frac{\sin 2a + i \sinh 2b}{\cos 2a + \cosh 2b}$$

$$= u + iv$$

Example:

$$\tan (2 + 3i) = -.00376 + 1.003i$$

LINE	DATA	OPERATIONS					DISPLAY	REMARKS
1	b	↑						
2	a	↑	2	x	STO	↑		
3		1	8	0	x	π		
4		÷	cos	x⇄y	2	x		
5		↑	↑	R↓	R↓	eˣ		
6		↑	¹/ₓ	+	2	÷		
7		+	RCL	x⇄y	STO	CLX		
8		+	x⇄y	eˣ	↑	¹/ₓ		
9		–	2	÷	RCL	÷		
10		x⇄y	1	8	0	x		
11		π	÷	sin	RCL	÷	u	
12		x⇄y					v	

Complex cotangent

Formula:

$$\cot (a + ib) = \frac{\sin 2a - i \sinh 2b}{\cosh 2a - \cos 2b}$$

$$= u + iv$$

Example:

$$\cot (2 + 3i) = -.0037 - .9968i$$

LINE	DATA	OPERATIONS					DISPLAY	REMARKS
1	b	↑						
2	a	↑	2	x	STO	↑		
3		1	8	0	x	π		
4		÷	cos	x⇄y	2	x		
5		↑	↑	R↓	R↓	eˣ		
6		↑	¹/ₓ	+	2	÷		
7		x⇄y	–	RCL	x⇄y	STO		
8		CLX	+	x⇄y	eˣ	↑		
9		¹/ₓ	–	2	÷	RCL		
10		÷	CHS	x⇄y	1	8		
11		0	x	π	÷	sin		
12		RCL	÷				u	
13		x⇄y					v	

Complex cosecant

Formula:

$$\csc (a + ib) = \frac{1}{\sin (a + ib)}$$

$$= u + iv$$

Example:

$$\csc (2 + 3i) = .09 + .0412i$$

LINE	DATA	OPERATIONS					DISPLAY	REMARKS
1	b	↑						
2	a	↑	1	8	0	x		
3		π	÷	STO	sin	x⇄y		
4		↑	↑	R↓	R↓	e^x		
5		↑	1/x	+	2	÷		
6		x	RCL	x⇄y	STO	CLX		
7		+	cos	x⇄y	e^x	↑		
8		1/x	−	2	÷	x		
9		RCL	↑	x	x⇄v	↑		
10		↑	x	RCL	R↓	x⇄y		
11		R↓	+	STO	÷		u	
12		x⇄y	RCL	÷	CHS		v	

Complex secant

Formula:

$$\sec (a + ib) = \frac{1}{\cos (a + ib)}$$

$$= u + iv$$

Example:

$$\sec (2 + 3i) = -.04 + .09i$$

LINE	DATA	OPERATIONS					DISPLAY	REMARKS
1	b	↑						
2	a	↑	1	8	0	x		
3		π	÷	STO	cos	x⇄y		
4		↑	↑	R↓	R↓	e^x		
5		↑	1/x	+	2	÷		
6		x	RCL	x⇄y	STO	CLX		
7		+	sin	x⇄y	e^x	↑		
8		1/x	−	2	÷	x		
9		CHS	RCL	↑	x	x⇄y		
10		↑	↑	x	RCL	R↓		
11		x⇄y	R↓	+	STO	÷	u	
12		x⇄y	RCL	÷	CHS		v	

356

Complex arc sine

Formula:

$$\text{arc sin } (a + ib) = \text{arc sin } \beta + i \text{ sgn}(b) \ln (\alpha + \sqrt{\alpha^2 - 1})$$

$$= u + iv$$

where: $\alpha = \frac{1}{2} \sqrt{(a + 1)^2 + b^2} + \frac{1}{2} \sqrt{(a - 1)^2 + b^2}$

$\beta = \frac{1}{2} \sqrt{(a + 1)^2 + b^2} - \frac{1}{2} \sqrt{(a - 1)^2 + b^2}$

$$\text{sgn } (b) = \begin{cases} 1 \text{ if } b \geqslant 0 \\ -1 \text{ if } b < 0. \end{cases}$$

Example:

$$\text{arc sin } (5 + 8i) = .556 + 2.9387i$$

Note: Inverse trigonometric and inverse hyperbolic functions are multiple-valued functions, but only one answer will be given for each of them.

LINE	DATA	OPERATIONS					DISPLAY	REMARKS
1	a	↑						
2	b	↑	x	STO	x⇄y	↑		
3		↑	1	+	↑	x		
4		x⇄y	1	−	↑	x		
5		RCL	+	√x	2	÷		
6		STO	R↓	+	√x	2		
7		÷	↑	↑	RCL	+		
8		x⇄y	RCL	−	arc	sin		
9		π	x	1	8	0		
10		÷					u	
11		R↓	↑	↑	x	1		
12		−	√x	+	ln			If b ⩾ 0, go to 14
13		CHS						
14							v	

Complex arc cosine

Formula:

$$\text{arc cos}(a + ib) = \text{arc cos } \beta - i \text{ sgn}(b) \ln\left(\alpha + \sqrt{\alpha^2 - 1}\right)$$

$$= u + iv$$

where: $\alpha = \dfrac{1}{2}\sqrt{(a+1)^2 + b^2} + \dfrac{1}{2}\sqrt{(a-1)^2 + b^2}$

$\beta = \dfrac{1}{2}\sqrt{(a+1)^2 + b^2} - \dfrac{1}{2}\sqrt{(a-1)^2 + b^2}$

$$\text{sgn}(b) = \begin{cases} 1 \text{ if } b \geqslant 0 \\ -1 \text{ if } b < 0 \end{cases}$$

Example:

$$\text{arc cos}(5 + 8i) = 1.0147 - 2.9387i$$

LINE	DATA	OPERATIONS						DISPLAY	REMARKS
1	a	↑							
2	b	↑	x	STO	x⇄y	↑			
3		↑	1	+	↑	x			
4		x⇄y	1	−	↑	x			
5		RCL	+	√x	2	÷			
6		STO	R↓	+	√x	2			
7		÷	↑	↑	RCL	+			
8		x⇄y	RCL	−	arc	cos			
9		π	x	1	8	0			
10		÷						u	
11		R↓	↑	↑	x	1			
12		−	√x	+	ln				If b < 0, go to 14
13		CHS							
14								v	

358

Complex arc tangent

Formula:

$$\text{arc tan}(a + ib) = \frac{1}{2}\left[\pi - \text{arc tan}\frac{1+b}{a} - \text{arc tan}\frac{1-b}{a}\right] + \frac{i}{4}\ln\left[\frac{(1+b)^2 + a^2}{(1-b)^2 + a^2}\right]$$

$$= u + iv$$

where: $(a + ib)^2 \neq -1$.

Example:

$$\text{arc tan}(5 + 8i) = 1.5142 + .0898i$$

LINE	DATA	OPERATIONS					DISPLAY	REMARKS
1	b	↑	↑	1	+			
2	a	STO	÷	arc	tan	x⇄y		
3		↑	1	x⇄y	–	↑		
4		↑	RCL	÷	arc	tan		
5		x⇄y	R↓	+	x⇄y	CLX		
6		1	8	0	x⇄y	–		
7		2	÷	π	x	1		
8		8	0	÷			u	
9		R↓	2	x⇄y	–	↑		
10		x	RCL	↑	x	STO		
11		+	x⇄y	↑	x	RCL		
12		+	÷	ln	4	÷	v	

Complex arc cotangent

Formula:

$$\text{arc cot}(a + ib) = \frac{\pi}{2} - \text{arc tan}(a + ib) = u + iv$$

Example:

$$\text{arc cot}(5 + 8i) = .0566 - .0898i$$

LINE	DATA	OPERATIONS					DISPLAY	REMARKS
1	b	↑	↑	1	+			
2	a	STO	÷	arc	tan	x⇄y		
3		↑	1	x⇄y	−	↑		
4		↑	RCL	÷	arc	tan		
5		x⇄y	R↓	+	x⇄y	CLX		
6		+	2	÷	π	x		
7		1	8	0	÷		u	
8		R↓	2	x⇄y	−	↑		
9		x	RCL	↑	x	STO		
10		+	x⇄y	↑	x	RCL		
11		+	÷	ln	4	÷		
12		CHS					v	

Complex arc cosecant

Formula:

$$\text{arc csc } (a + ib) = \text{arc sin}\left(\frac{1}{a + ib}\right)$$

$$= u + iv$$

Example:

$$\text{arc csc } (5 + 8i) = .05598 - .0899i$$

$$(D < 0)$$

LINE	DATA	OPERATIONS					DISPLAY	REMARKS
1	b	↑						
2	a	↑	↑	x	STO	CLX		
3		+	R↓	x	RCL	+		
4		STO	÷	CHS	x⇄y	RCL		
5		÷	x⇄y				D	
6		↑	x	STO	x⇄y	↑		
7		↑	1	+	↑	x		
8		x⇄y	1	−	↑	x		
9		RCL	+	√x	2	÷		
10		STO	R↓	+	√x	2		
11		÷	↑	↑	RCL	+		
12		x⇄y	RCL	−	arc	sin		
13		π	x	1	8	0		
14		÷					u	
15		R↓	↑	↑	x	1		
16		−	√x	+	ln			If D > 0, go to 18
17		CHS						
18							v	

361

Complex arc secant

Formula:

$$\text{arc sec}(a + ib) = \text{arc cos}\left(\frac{1}{a + ib}\right)$$

$$= u + iv$$

Example:

$$\text{arc sec}(5 + 8i) = 1.5148 + .0899i$$

$$(D < 0)$$

LINE	DATA	OPERATIONS					DISPLAY	REMARKS
1	b	↑						
2	a	↑	↑	x	STO	CLX		
3		+	R↓	x	RCL	+		
4		STO	÷	CHS	x⇄y	RCL		
5		÷	x⇄y				D	
6		↑	x	STO	x⇄y	↑		
7		↑	1	+	↑	x		
8		x⇄y	1	−	↑	x		
9		RCL	+	√x	2	÷		
10		STO	R↓	+	√x	2		
11		÷	↑	↑	RCL	+		
12		x⇄y	RCL	−	arc	cos		
13		π	x	1	8	0		
14		÷					u	
15		R↓	↑	↑	x	1		
16		−	√x	+	ln			If D < 0, go to 18
17		CHS						
18							v	

362

Complex hyperbolic sine

Formula:

$$\sinh(a + ib) = -i \sin i (a + ib)$$

$$= u + iv$$

Example:

$$\sinh(3 - 2i) = -4.1689 - 9.154i$$

LINE	DATA	OPERATIONS					DISPLAY	REMARKS
1	a	↑						
2	b	CHS	↑	1	8	0		
3		x	π	÷	STO	sin		
4		x⇄y	↑	↑	R↓	R↓		
5		eˣ	↑	1/x	+	2		
6		÷	x	RCL	x⇄y	STO		
7		CLX	+	cos	x⇄y	eˣ		
8		↑	1/x	−	2	÷		
9		x	RCL	CHS	x⇄y		u	
10		x⇄y					v	

Complex hyperbolic cosine

Formula:

$$\cosh(a + ib) = \cos i (a + ib)$$

$$= u + iv$$

Example:

$$\cosh(1 + 2i) = -.6421 + 1.0686i$$

363

LINE	DATA	OPERATIONS					DISPLAY	REMARKS
1	a	↑						
2	b	CHS	↑	1	8	0		
3		x	π	÷	STO	cos		
4		x⇄y	↑	↑	R↓	R↓		
5		e^x	↑	1/x	+	2		
6		÷	x	RCL	x⇄y	STO	u	
7		CLX	+	sin	x⇄y	e^x		
8		↑	1/x	−	2	÷		
9		x	CHS				v	

Complex hyperbolic tangent

Formula:

$$\tanh(a + ib) = -\,i\,\tan\,i\,(a + ib)$$

$$= u + iv$$

Example:

$$\tanh(1 + 2i) = 1.1667 - .243i$$

LINE	DATA	OPERATIONS					DISPLAY	REMARKS
1	a	↑						
2	b	CHS	↑	2	x	STO		
3		↑	1	8	0	x		
4		π	÷	cos	x⇄y	2		
5		x	↑	↑	R↓	R↓		
6		e^x	↑	1/x	+	2		
7		÷	+	RCL	x⇄y	STO		
8		CLX	+	x⇄y	e^x	↑		
9		1/x	−	2	÷	RCL		
10		÷	x⇄y	1	8	0		
11		x	π	÷	sin	RCL		
12		÷	CHS	x⇄y			u	
13		x⇄y					v	

Complex hyperbolic cotangent

Formula:

$$\coth (a + ib) = i \cot i (a + ib)$$

$$= u + iv$$

Example:

$$\coth (1 + 2i) = .8213 + .17138i$$

LINE	DATA	OPERATIONS					DISPLAY	REMARKS
1	a	↑						
2	b	CHS	↑	2	x	STO		
3		↑	1	8	0	x		
4		π	÷	cos	x⇄y	2		
5		x	↑	↑	R↓	R↓		
6		eˣ	↑	¹/ₓ	+	2		
7		÷	x⇄y	−	RCL	x⇄y		
8		STO	CLX	+	x⇄y	eˣ		
9		↑	¹/ₓ	−	2	÷		
10		RCL	÷	CHS	x⇄y	1		
11		8	0	x	π	÷		
12		sin	RCL	÷	x⇄y	CHS	u	
13		x⇄y					v	

Complex hyperbolic cosecant

Formula:

$$\operatorname{csch} (a + ib) = i \csc i (a + ib)$$

$$= u + iv$$

Example:

$$\operatorname{csch} (1 + 2i) = -.2215 - .63549i$$

LINE	DATA	OPERATIONS					DISPLAY	REMARKS
1	a	↑						
2	b	CHS	↑	1	8	0		
3		x	π	÷	STO	sin		
4		x⇄y	↑	↑	R↓	R↓		
5		e^x	↑	$1/x$	+	2		
6		÷	x	RCL	x⇄y	STO		
7		CLX	+	cos	x⇄y	e^x		
8		↑	$1/x$	−	2	÷		
9		x	RCL	↑	x	x⇄y		
10		↑	↑	x	RCL	R↓		
11		x⇄y	R↓	+	STO	÷		
12		x⇄y	RCL	÷			u	
13		x⇄y					v	

Complex hyperbolic secant

Formula:

$$\text{sech} \, (a + ib) = \sec i \, (a + ib)$$

$$= u + iv$$

Example:

$$\text{sech} \, (1 + 2i) = -.4131 - .6875i$$

LINE	DATA	OPERATIONS					DISPLAY	REMARKS
1	a	↑						
2	b	CHS	↑	1	8	0		
3		x	π	÷	STO	cos		
4		x⇄y	↑	↑	R↓	R↓		
5		e^x	↑	$1/x$	+	2		
6		÷	x	RCL	x⇄y	STO		
7		CLX	+	sin	x⇄y	e^x		
8		↑	$1/x$	−	2	÷		
9		x	CHS	RCL	↑	x		
10		x⇄y	↑	↑	x	RCL		
11		R↓	x⇄y	R↓	+	STO		
12		÷					u	
13		x⇄y	RCL	÷	CHS		v	

Complex inverse hyperbolic sine

Formula:

$$\sinh^{-1}(a + ib) = -i \text{ arc sin } i(a + ib)$$

$$= u + iv$$

Example:

$$\sinh^{-1}(8 - 5i) = 2.9387 - .556i$$

LINE	DATA	OPERATIONS					DISPLAY	REMARKS
1	a	↑						
2	b	CHS	x⇄y	↑	x	STO		
3		x⇄y	↑	↑	1	+		
4		↑	x	x⇄y	1	–		
5		↑	x	RCL	+	√x̄		
6		2	÷	STO	R↓	+		
7		√x̄	2	÷	↑	↑		
8		RCL	+	x⇄y	RCL	–		
9		arc	sin	π	x	1		
10		8	0	÷	CHS	STO		
11		R↓	↑	↑	x	1		
12		–	√x̄	+	ln			If a ≥ 0, go to 14
13		CHS						
14							u	
15		RCL					v	

Complex inverse hyperbolic cosine

Formula:

$$\cosh^{-1}(a + ib) = i \text{ arc cos }(a + ib)$$

$$= u + iv$$

Example:

$$\cosh^{-1}(5 + 8i) = 2.9387 + 1.0147i$$

LINE	DATA	OPERATIONS					DISPLAY	REMARKS
1	a	↑						
2	b	↑	x	STO	x⇄y	↑		
3		↑	1	+	↑'	x		
4		x⇄y	1	−	↑	x		
5		RCL	+	√x	2	÷		
6		STO	R↓	+	√x	2		
7		÷	↑	↑	RCL	+		
8		x⇄y	RCL	−	arc	cos		
9		π	x	1	8	0		
10		÷	STO	R↓	↑	↑		
11		x	1	−	√x	+		
12		ln						If b ≥ 0, go to 14
13		CHS						
14							u	
15		RCL					v	

Complex inverse hyperbolic tangent

Formula:

$$\tanh^{-1}(a + ib) = -i \, \text{arc tan} \, i(a + ib)$$

$$= u + iv$$

Example:

$$\tanh^{-1}(8 - 5i) = .0898 - 1.5142i$$

LINE	DATA	OPERATIONS					DISPLAY	REMARKS
1	a	↑	↑	1	+			
2	b	CHS	STO	÷	arc	tan		
3		x⇄y	↑	1	x⇄y	−		
4		↑	↑	RCL	÷	arc		
5		tan	x⇄y	R↓	+	x⇄y		
6		CLX	1	8	0	x⇄y		
7		−	2	÷	π	x		
8		1	8	0	÷	CHS	v	
9		R↓	2	x⇄y	−	↑		
10		x	RCL	↑	x	STO		
11		+	x⇄y	↑	x	RCL		
12		+	÷	ln	4	÷	u	

Complex inverse hyperbolic cotangent

Formula:

$$\coth^{-1}(a+ib) = i \text{ arc cot } i(a+ib)$$

$$= u + iv$$

Example:

$$\coth^{-1}(8-5i) = .0898 + .0566i$$

LINE	DATA	OPERATIONS					DISPLAY	REMARKS
1	a	↑	↑	1	+			
2	b	CHS	STO	÷	arc	tan		
3		x⇄y	↑	1	x⇄y	−		
4		↑	↑	RCL	÷	arc		
5		tan	x⇄y	R↓	+	x⇄y		
6		CLX	+	2	÷	π		
7		x	1	8	0	÷	v	
8		R↓	2	x⇄y	−	↑		
9		x	RCL	↑	x	STO		
10		+	x⇄y	↑	x	RCL		
11		+	÷	ln	4	÷	u	

Complex inverse hyperbolic cosecant

Formula:

$$\mathrm{csch}^{-1}\,(a + ib) = i\ \mathrm{arc}\ \mathrm{csc}\ i\,(a + ib)$$

$$= u + iv$$

Example:

$$\mathrm{csch}^{-1}\,(8 - 5i) = .0899 + .05598i$$

$$(D < 0)$$

LINE	DATA	OPERATIONS					DISPLAY	REMARKS
1	a	↑						
2	b	CHS	↑	↑	x	STO		
3		CLX	+	R↓	x	RCL		
4		+	STO	÷	CHS	x⇄y		
5		RCL	÷	x⇄y			D	
6		↑	x	STO	x⇄y	↑		
7		↑	1	+	↑	x		
8		x⇄y	1	−	↑	x		
9		RCL	+	√x	2	÷		
10		STO	R↓	+	√x	2		
11		÷	↑	↑	RCL	+		
12		x⇄y	RCL	−	arc	sin		
13		π	x	1	8	0		
14		÷	STO	R↓	↑	↑		
15		x	1	−	√x	+		
16		ln						If D < 0, go to 18.
17		CHS						
18							u	
19		RCL					v	

Complex inverse hyperbolic cosecant

Formula:

$$\operatorname{csch}^{-1}(a + ib) = i \operatorname{arc\,csc} i (a + ib)$$

$$= u + iv$$

Example:

$$\operatorname{csch}^{-1}(8 - 5i) = .0899 + .05598i$$

$$(D < 0)$$

LINE	DATA	OPERATIONS					DISPLAY	REMARKS
1	a	↑						
2	b	CHS	↑	↑	x	STO		
3		CLX	+	R↓	x	RCL		
4		+	STO	÷	CHS	x⇄y		
5		RCL	÷	x⇄y			D	
6		↑	x	STO	x⇄y	↑		
7		↑	1	+	↑	x		
8		x⇄y	1	−	↑	x		
9		RCL	+	√x̄	2	÷		
10		STO	R↓	+	√x̄	2		
11		÷	↑	↑	RCL	+		
12		x⇄y	RCL	−	arc	sin		
13		π	x	1	8	0		
14		÷	STO	R↓	↑	↑		
15		x	1	−	√x̄	+		
16		ln						If D < 0, go to 18.
17		CHS						
18							u	
19		RCL					v	

Compound Interest
See page 96

Hyperbolic and Inverse Hyperbolic Functions

Register usage for the following functions:

(Same as for keyboard trigonometric functions.)

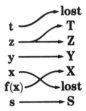

Gudermannian function

Formula:

$$\text{gd } x = 2 \text{ arc tan } e^x - \frac{\pi}{2}$$

Example:

$$\text{gd } 0.345 = 19.386$$

Note: $\frac{\pi}{2} = 90°$

LINE	DATA	OPERATIONS					DISPLAY	REMARKS
1	x	e^x	arc	tan	2	x		
2		9	0	−				

Hyperbolic sine

Formula:

$$\sinh x = \frac{e^x - e^{-x}}{2}$$

Example:

$$\sinh 3.2 = 12.25$$

LINE	DATA	OPERATIONS					DISPLAY	REMARKS
1	x	e^x	↑	$1/x$	−	2		
2		÷						

Hyperbolic cosine

Formula:

$$\cosh x = \frac{e^x + e^{-x}}{2}$$

Example:

$$\cosh 3.2 = 12.29$$

LINE	DATA	OPERATIONS					DISPLAY	REMARKS
1	x	e^x	↑	¹/ₓ	+	2		
2		÷						

Hyperbolic tangent

Formula:

$$\tanh x = \frac{\sinh x}{\cosh x} = \sin \mathrm{gd}\, x$$

Example:

$$\tanh 3.2 = .99668$$

LINE	DATA	OPERATIONS					DISPLAY	REMARKS
1	x	e^x	arc	tan	2	x		
2		9	0	−	sin			

Hyperbolic cotangent

Formula:

$$\coth x = \frac{1}{\tanh x}$$

Example:

$$\coth 3.2 = 1.003$$

LINE	DATA	OPERATIONS					DISPLAY	REMARKS
1	x	e^x	arc	tan	2	x		
2		9	0	−	sin	¹/ₓ		

Hyperbolic cosecant

Formula:

$$\text{csch } x = \frac{1}{\sinh x}$$

Example:

$$\text{csch } 3.2 = .08166$$

LINE	DATA	OPERATIONS					DISPLAY	REMARKS
1	x	e^x	↑	¹/ₓ	−	2		
2		÷	¹/ₓ					

Hyperbolic secant

Formula:

$$\text{sech } x = \frac{1}{\cosh x}$$

Example:

$$\text{sech } 3.2 = .081$$

LINE	DATA	OPERATIONS					DISPLAY	REMARKS
1	x	e^x	↑	¹/ₓ	+	2		
2		÷	¹/ₓ					

Inverse Gudermannian function

Formula:

$$\text{gd}^{-1}\theta = \ln \tan\left(\frac{\pi}{4} + \frac{\theta}{2}\right)$$

Example:

$$\text{gd}^{-1} \ 30° = .549$$

Note: $\dfrac{\pi}{4} = 45°$

LINE	DATA	OPERATIONS					DISPLAY	REMARKS
1	θ	↑	2	÷	4	5		
2		+	tan	ln				

Inverse hyperbolic sine

Formula:

$$\sinh^{-1} x = \ln\left[x + (x^2 + 1)^{\frac{1}{2}}\right]$$
$$= gd^{-1}\,(\tan^{-1} x)$$

Example:

$$\sinh^{-1} 51.777 = 4.64$$

LINE	DATA	OPERATIONS					DISPLAY	REMARKS
1	x	arc	tan	2	÷	4		
2		5	+	tan	ln			

Inverse hyperbolic tangent

Formula:

$$\tanh^{-1} x = \frac{1}{2}\ln\frac{1+x}{1-x} = gd^{-1}\,(\sin^{-1} x)$$

Example:

$$\tanh^{-1} 0.777 = 1.038$$

LINE	DATA	OPERATIONS					DISPLAY	REMARKS
1	x	arc	sin	2	÷	4		
2		5	+	tan	ln			

Inverse hyperbolic secant

Formula:

$$\text{sech}^{-1} x = \ln\left[\frac{1}{x} + \left(\frac{1}{x^2} - 1\right)^{\frac{1}{2}}\right] = gd^{-1}\,(\cos^{-1} x)$$

Example:

$$\text{sech}^{-1} 0.777 = .74$$

LINE	DATA	OPERATIONS					DISPLAY	REMARKS
1	x	arc	cos	2	÷	4		
2		5	+	tan	ln			

Inverse hyperbolic cosine

Formula:

$$\cosh^{-1} x = \operatorname{sech}^{-1} \frac{1}{x}$$

Example:

$$\cosh^{-1} 51.777 = 4.64$$

LINE	DATA	OPERATIONS					DISPLAY	REMARKS
1	x	$^1/_x$	arc	cos	2	÷		
2		4	5	+	tan	ln		

Inverse hyperbolic cotangent

Formula:

$$\coth^{-1} x = \tanh^{-1} \frac{1}{x}$$

Example:

$$\coth^{-1} 51.777 = .0193$$

LINE	DATA	OPERATIONS					DISPLAY	REMARKS
1	x	$^1/_x$	arc	sin	2	÷		
2		4	5	+	tan	ln		

Inverse hyperbolic cosecant

Formula:

$$\operatorname{csch}^{-1} x = \sinh^{-1} \frac{1}{x}$$

Example:

$$\operatorname{csch}^{-1} 0.777 = 1.0705$$

LINE	DATA	OPERATIONS					DISPLAY	REMARKS
1	x	$^1/_x$	arc	tan	2	÷		
2		4	5	+	tan	ln		

INDEX